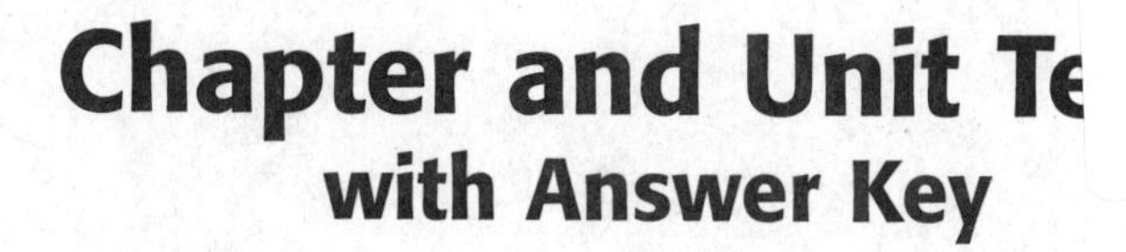

HOLT

World Geography Today

HOLT, RINEHART AND WINSTON
A Harcourt Education Company
Austin • New York • Orlando • Atlanta • San Francisco • Boston • Dallas • Toronto • London

Cover description: photograph showing spear fishers in the Cook Islands of the South Pacific

Cover credit: Nicholas DeVore/Getty Images/Stone

Printed in the United States of America

ISBN 0-03-038867-8

2 3 4 5 6 7 8 9 095 07 06 05

Contents

Chapter and Unit Tests

Unit 1 The Geographer's World

Unit 2 The United States and Canada

Unit 3 Middle and South America

Unit 4 Europe

Unit 5 Russia and Northern Eurasia

Name ______________________ Class ______________ Date ______________

Chapter Test Form A

Studying Geography

REVIEWING FACTS

MATCHING *(3 points each)* In the space provided, write the letter that matches each description. Choose your answers from the list below. Some answers will not be used.

_____ **1.** The study of weather

_____ **2.** The Great Lakes

_____ **3.** Lines drawn on the globe in an east-west direction

_____ **4.** A collection of maps in a book

_____ **5.** The physical, human, and cultural features of a place

_____ **6.** The elevation, layout, and shape of the land

_____ **7.** The airports of the United States

_____ **8.** The unit of measurement for distance between parallels or meridians

_____ **9.** When all of the territories within a country are connected

_____ **10.** The way a person looks at something

a. landscape
b. topography
c. meteorology
d. functional region
e. latitude
f. formal region
g. atlas
h. longitude
i. perspective
j. cartography
k. contiguous
l. degree

FILL IN THE BLANK *(3 points each)* For each of the following statements, fill in the blank with the appropriate word, phrase, or place.

1. The ______________________ divides Earth into Northern and Southern Hemispheres.

2. The two main branches of geography are ______________________ geography and physical geography.

3. The features and symbols of a map are described in the ______________________.

4. The Atlantic, Pacific, Arctic, and ______________________ Oceans make up the global ocean.

5. A ______________________ is a directional indicator with arrows that point in all four directions.

6. The theme of ______________________ covers how people and things change location and the effects of these changes.

7. A ______________________ map will give information on the distribution of people in a region.

8. The shortest route between any two places on the planet is called a

______________________ route.

9. A ______________________ map uses lines to connect points of equal elevation above or below sea level.

10. A ______________________ projection shows true area sizes, but it distorts shapes.

UNDERSTANDING IDEAS *(3 points each)* For each of the following, write the letter of the best choice in the space provided.

______ **1.** Which of the following is not a common map element?
- **a.** key
- **b.** distance scale
- **c.** contour lines
- **d.** directional indicator

______ **2.** The study of maps and mapmaking is called
- **a.** perception.
- **b.** perspective.
- **c.** cartography.
- **d.** meteorology.

______ **3.** Another name for a population pyramid is
- **a.** population grid.
- **b.** age-structure diagram.
- **c.** age-ratio grid.
- **d.** age-climate graph.

______ **4.** Which of the following is not a theme of geography?
- **a.** human-environment interaction
- **b.** perceptual region
- **c.** place
- **d.** location

______ **5.** Which essential element focuses on the study of earthquakes, volcanoes, and weather patterns?
- **a.** places and regions
- **b.** physical systems
- **c.** the world in spatial terms
- **d.** the uses of geography

PRACTICING SKILLS *(5 points each)* Study the contour map and answer the questions that follow.

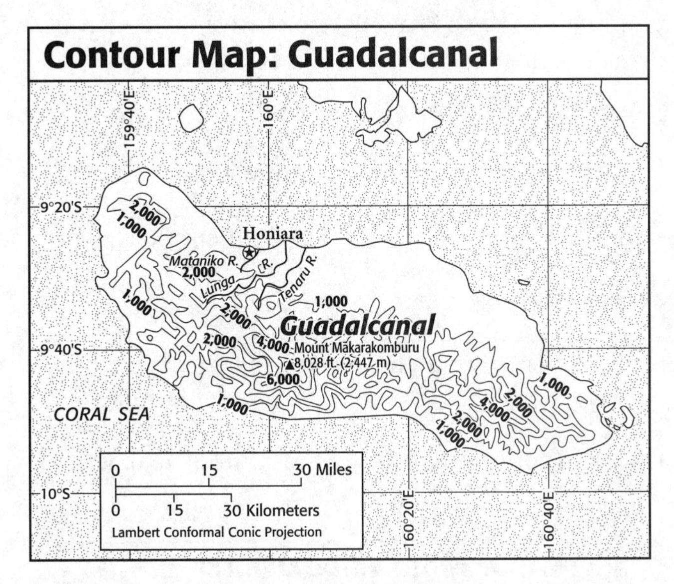

1. What is the approximate longitude and latitude of the capital city of Guadalcanal?

____

____

2. What is the elevation in the area of the Lunga River?

____

____

COMPOSING AN ESSAY *(15 points)* On a separate sheet of paper, write an essay in response to *one* of the following.

1. Describe the three main types of map projections and the advantages and disadvantages of each type.

2. Discuss why geography and cartography are essential in today's world.

Name ______________________ Class ______________ Date ______________

Chapter Test Form B

Studying Geography

SHORT ANSWER *(10 points each)* Provide brief answers for each of the following. Use examples to support your answers.

1. List the five themes of geography.

2. List the three types of regions and give one example of each.

3. What are three industries or businesses that might employ geographers?

4. Describe three special purpose maps.

5. Describe how geographers have made it possible to accurately locate places on Earth.

PRACTICING SKILLS *(10 points each)* Study the contour map and answer the questions that follow.

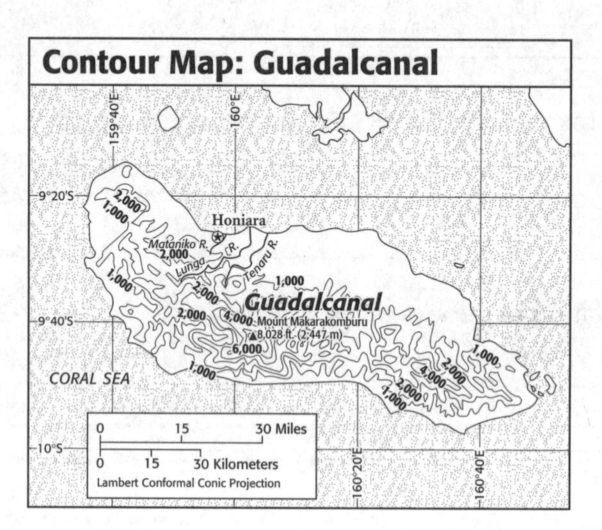

1. Based on this map, describe Guadalcanal.

2. Where is the sharpest change in elevation?

COMPOSING AN ESSAY *(30 points)* On a separate sheet of paper, write an essay in response to *one* of the following.

1. How does the study of human geography differ from the study of physical geography?

2. Describe the six essential elements and how they relate to the five themes of geography.

3. Compare and contrast the three types of map projections.

4. How have advances in computers and satellite pictures affected cartography?

Name ______________________ Class ______________ Date ______________

Chapter Test Form A

Earth in Space

REVIEWING FACTS

MATCHING *(3 points each)* In the space provided, write the letter that matches each description. Choose your answers from the list below. Some answers will not be used.

______ **1.** The part of Earth that includes all life forms

______ **2.** Any body that orbits a larger body

______ **3.** Warm low-latitude areas near the equator

______ **4.** One complete spin of Earth on its axis

______ **5.** Time when Earth's poles are not pointed toward or away from the Sun

______ **6.** All the biological, chemical, and physical conditions that interact and affect life

______ **7.** One elliptical orbit of Earth around the Sun

______ **8.** Energy from the Sun

______ **9.** Time that Earth's poles point at their greatest angle toward or away from the Sun

______**10.** The solid crust of the planet

a. equinox
b. rotation
c. lithosphere
d. hydrosphere
e. satellite
f. tilt
g. solstice
h. biosphere
i. revolution
j. tropics
k. environment
l. solar energy

FILL IN THE BLANK *(3 points each)* For each of the following statements, fill in the blank with the appropriate word, phrase, or place.

1. The planets are visible because they ______________________ sunlight.

2. In the Northern Hemisphere, the March equinox marks the beginning of

______________________.

3. Earth is different from all other planets in the solar system because it has

______________________.

4. The Sun is about ______________________ times larger than Earth.

5. The shape of Earth's orbit around the Sun is ______________________.

6. The ocean floor is part of the ______________________.

7. The North Pole points toward the ______________________.

8. Earth's ______________________ holds the atmosphere around the planet.

9. Earth revolves around the Sun in ______________________ days.

10. ______________________ are caused by the gravitational pull of the Sun, Moon, and Earth.

UNDERSTANDING IDEAS *(3 points each)* For each of the following, write the letter of the best choice in the space provided.

______ **1.** The Tropic of Cancer is located
- **a.** in the Southern Hemisphere.
- **b.** south of the Equator.
- **c.** in the Northern Hemisphere.
- **d.** north of the Arctic Circle.

______ **2.** The Milky Way is a
- **a.** solar system.
- **b.** satellite
- **c.** planet.
- **d.** galaxy.

______ **3.** What causes the Sun to appear to rise and set?
- **a.** Earth's tilt
- **b.** Earth's revolution
- **c.** Earth's rotation
- **d.** Earth's satellite

______ **4.** Which sphere contains clouds and fog?
- **a.** biosphere
- **b.** atmosphere
- **c.** lithosphere
- **d.** hydrosphere

______ **5.** Which is not an inner planet?
- **a.** Mercury
- **b.** Venus
- **c.** Saturn
- **d.** Earth

Name ______________________ Class ______________ Date ____________

PRACTICING SKILLS *(5 points each)* Study the diagram and answer the questions that follow.

The Seasons

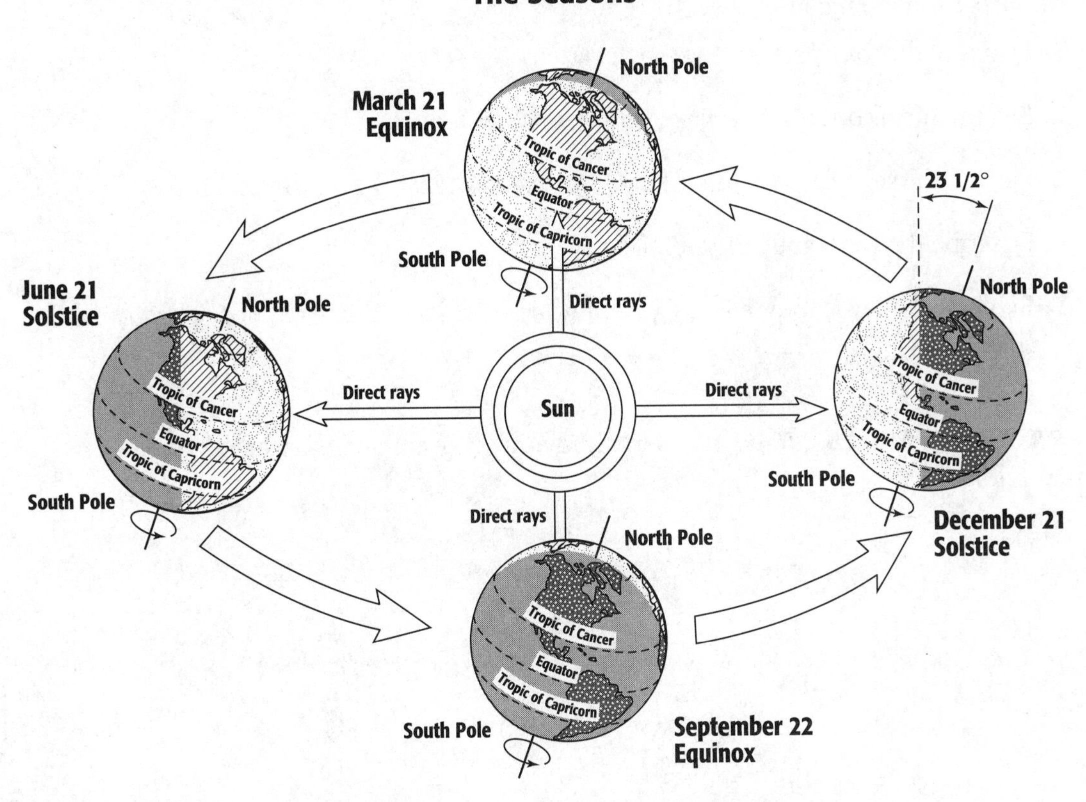

1. How does the tilt of Earth's axis affect the seasons?

____

____

2. On which day is the South Pole pointed away from the Sun?

____

____

COMPOSING AN ESSAY *(15 points)* On a separate sheet of paper, write an essay in response to *one* of the following.

1. Explain why knowledge of the four spheres of the environment is important to the quality of human life.

2. Describe life on Earth if Earth revolved around the Sun but did not rotate on its axis.

Name ______________________ Class ______________ Date ____________

Chapter Test Form B

Earth in Space

SHORT ANSWER *(10 points each)* Provide brief answers for each of the following. Use examples to support your answers.

1. Describe the four parts of the Earth system.
2. List the major bodies that make up the solar system.
3. During the solstices, where do Earth's poles point in relation to the Sun?
4. Describe the geography of the Moon.
5. Explain why low-latitude areas generally are warm and polar regions are generally cold.

PRACTICING SKILLS *(10 points each)* Study the diagram and answer the questions that follow.

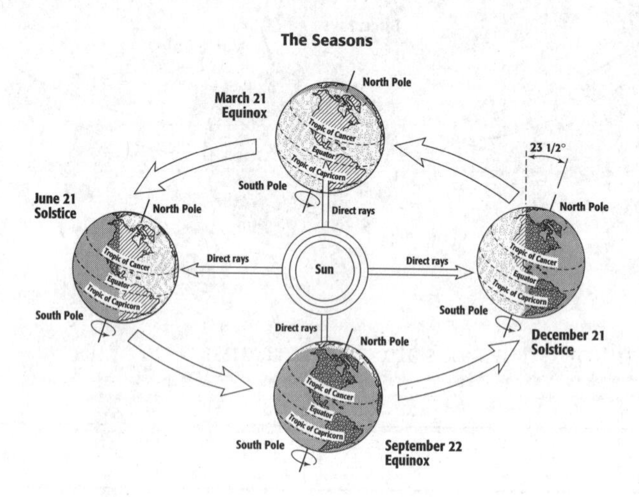

1. What relationship between Earth and the Sun causes the seasons?
2. During which season does the Arctic Circle receive the most sunlight?

COMPOSING AN ESSAY *(30 points)* On a separate sheet of paper, write an essay in response to *one* of the following.

1. Describe the three factors that affect how much solar energy is received by various places on Earth.
2. Compare and contrast solstices and equinoxes.
3. How do the inner planets and the outer planets differ?
4. Describe the organization of the universe.

Name ______________________ Class ______________ Date ____________

Chapter Test Form A

Weather and Climate

REVIEWING FACTS

MATCHING *(3 points each)* In the space provided, write the letter that matches each description. Choose your answers from the list below. Some answers will not be used.

_____ **1.** The meeting of two air masses of widely different temperatures or moisture levels

_____ **2.** The amount of water vapor in the air

_____ **3.** The weather conditions in a geographic area over a long period of time

_____ **4.** A community of living and nonliving organisms

_____ **5.** The drier, leeward side of a mountain

_____ **6.** Any center of low pressure

_____ **7.** Dry

_____ **8.** The increasing temperature of Earth over the last few decades

_____ **9.** Powerful rotating storms in the western Pacific Ocean

_____ **10.** Soil that remains frozen year-round

a. global warming
b. orographic effect
c. ecosystem
d. condensation
e. humidity
f. cyclone
g. arid
h. permafrost
i. front
j. rain shadow
k. climate
l. typhoons

FILL IN THE BLANK *(3 points each)* For each of the following statements, fill in the blank with the appropriate word, phrase, or place.

1. A ______________________ area occurs when air is heated, expands, becomes less dense, and rises.

2. Fog is one form of ______________________.

3. ______________________ are twisting spirals of air.

4. ______________________ is the horizontal flow of air.

5. Dry climates have low annual rainfall, but their ______________________ can vary greatly.

6. Ocean ______________________ are caused by wind, Earth's rotation, and varying ocean temperature.

7. Both deciduous and coniferous forests are found in the ______________________ climates.

8. The trade winds are an example of ______________________ winds.

9. The climate most influenced by the oceans is the ______________________ climate.

10. You could expect to experience the most precipitation in persistent

______________________-pressure zones.

UNDERSTANDING IDEAS *(3 points each)* For each of the following, write the letter of the best choice in the space provided.

______ **1.** Mountains can influence climate through
- **a.** the orographic effect.
- **b.** global warming.
- **c.** the greenhouse effect.
- **d.** humidity.

______ **2.** The condition of the atmosphere at a given time and place is called
- **a.** temperature.
- **b.** climate.
- **c.** weather.
- **d.** greenhouse effect.

______ **3.** Where are the most diverse ecosystems found?
- **a.** in dry climates
- **b.** in high latitude climates
- **c.** in tropical climate regions
- **d.** in middle-latitude climates

______ **4.** Which of the following appears to be a factor in global warming?
- **a.** melting polar ice caps
- **b.** increased levels of carbon dioxide
- **c.** a mini ice age
- **d.** increased levels of oxygen

______ **5.** What causes air pressure to decrease?
- **a.** increased oxygen levels
- **b.** increased temperature
- **c.** decreased temperature
- **d.** decreased carbon dioxide levels

Name ______________________ Class ______________ Date ______________

PRACTICING SKILLS *(5 points each)* Study the diagram and answer the questions that follow.

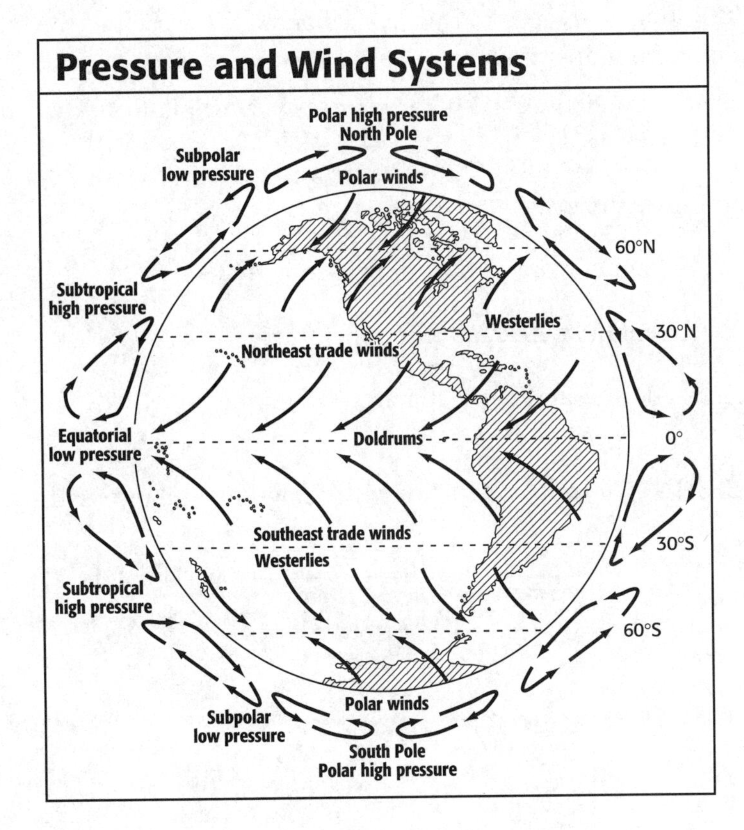

1. Why do the areas alternate between high and low pressure?

____

____

2. In which area is most of the United States located?

____

____

COMPOSING AN ESSAY *(15 points)* On a separate sheet of paper, write an essay in response to *one* of the following.

1. Describe the global wind belts and how they interact with climate.

2. Describe the physical factors of Earth that affect climate.

Name ____________________ Class ____________ Date ____________

Chapter Test Form B

Weather and Climate

SHORT ANSWER *(10 points each)* Provide brief answers for each of the following. Use examples to support your answers.

1. List four forms of condensation and four forms of precipitation. How are they different?

2. What factors create storms?

3. Why are the different climates grouped by latitude?

4. How does the greenhouse effect affect temperature?

5. How do oceans affect weather and climate?

PRACTICING SKILLS *(10 points each)* Study the diagram and answer the questions that follow.

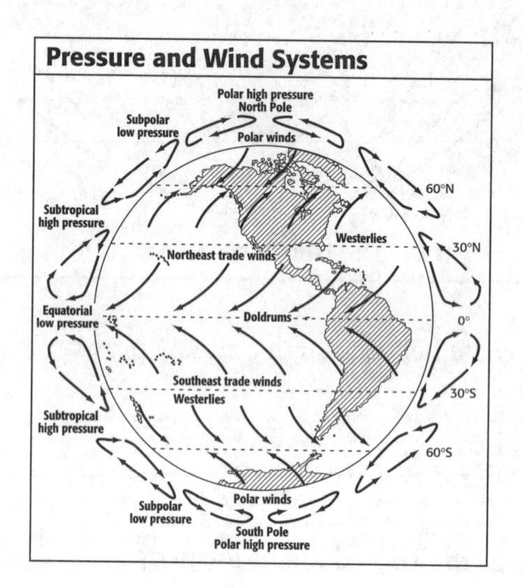

1. Why do the trade winds not flow directly north and south?

2. Why is the air pressure in the subpolar regions low?

COMPOSING AN ESSAY *(30 points)* On a separate sheet of paper, write an essay in response to *one* of the following.

1. How do the global wind belts affect human life?

2. How do mountain ranges within the United States affect its climate?

3. Around which latitudes are most tropical climates located? Why?

4. Compare and contrast the tropical humid climate and the tropical wet and dry climate.

Name ______________________ Class ____________ Date ____________

Chapter Test Form A

Landforms, Water, and Natural Resources

REVIEWING FACTS

MATCHING *(3 points each)* In the space provided, write the letter that matches each description. Choose your answers from the list below. Some answers will not be used.

_____ **1.** Water vapor containing chemicals from air pollution

_____ **2.** The level below ground at which all the spaces inside rock are filled with water

_____ **3.** Salt buildup in the soil

_____ **4.** The downward movement of minerals and humus in soil

_____ **5.** A stream or river that flows into a larger stream or river

_____ **6.** Earth's center

_____ **7.** The movement of surface material from one place to another

_____ **8.** Liquid rock within Earth

_____ **9.** Broken down plant and animal material

_____ **10.** An area drained by a river and its tributaries

a. soil salinization
b. leaching
c. weathering
d. acid rain
e. water table
f. magma
g. watershed
h. erosion
i. tributary
j. core
k. humus
l. headwaters

FILL IN THE BLANK *(3 points each)* For each of the following statements, fill in the blank with the appropriate word, phrase, or place.

1. Gravel, sand, and mud are all forms of ______________________.

2. Most of Earth's mass is found in the ______________________.

3. Soil is a ______________________ resource.

4. The theory of ______________________ explains how forces within the planet create landforms.

5. A watershed begins with ______________________.

6. When mined, minerals are removed from the ground in the form of

______________________.

Name ______________________ Class ______________ Date ______________

7. ______________________ occurs where two plates move laterally past each other.

8. Contour plowing is one way to reduce soil ______________________.

9. Most agriculture uses the ______________________ of the soil horizon.

10. Petrochemicals come from ______________________.

UNDERSTANDING IDEAS *(3 points each)* For each of the following, write the letter of the best choice in the space provided.

______ **1.** What three processes create landforms?
- **a.** tectonic processes, erosion, and weathering
- **b.** tectonic processes, erosion, and sediment
- **c.** erosion, sediment, and weathering
- **d.** sediment, tectonic processes, and weathering

______ **2.** One result of a lowered water table is
- **a.** ground settlement or slumping.
- **b.** decreased wetlands.
- **c.** flooding.
- **d.** soil erosion.

______ **3.** The soil horizon is composed of
- **a.** humus, subsoil, and weathered rock.
- **b.** humus, groundwater, and the aquifer.
- **c.** humus, subsoil, and groundwater.
- **d.** humus, groundwater, and weathered rock.

______ **4.** You might find an aquifer
- **a.** in a dam.
- **b.** near the ocean.
- **c.** near the groundwater.
- **d.** in a reservoir.

______ **5.** The deep valley marking a plate collision is called a
- **a.** rift.
- **b.** trench.
- **c.** fault.
- **d.** fold.

Name ______________________ Class ______________ Date ______________

PRACTICING SKILLS *(5 points each)* Study the diagram and answer the questions that follow.

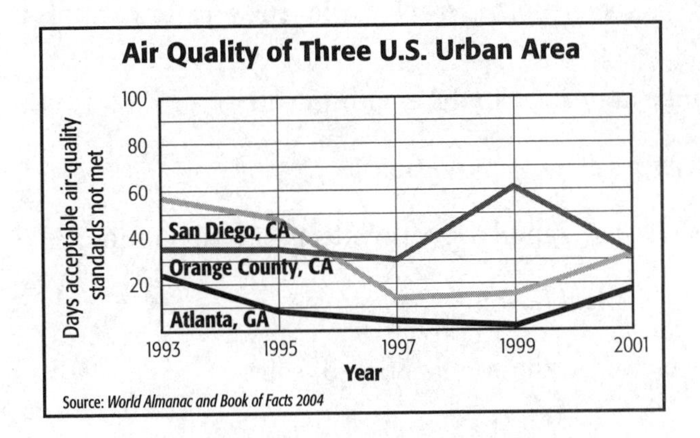

1. How many days did Orange County fail to meet acceptable air quality standards in 1999?

____

____

2. Which location had the greatest reduction in the number of days of unacceptable air quality between 1995 and 1997?

____

____

COMPOSING AN ESSAY *(15 points)* On a separate sheet of paper, write an essay in response to *one* of the following.

1. What might be the effect of increased population on Earth's natural resources?

2. Describe how weathering affects the environment.

Name ____________________ Class ____________ Date ____________

Chapter Test Form B

Landforms, Water, and Natural Resources

SHORT ANSWER *(10 points each)* Provide brief answers for each of the following. Use examples to support your answers.

1. Why are wetlands important to the environment?

2. What are three types of plate movement?

3. What are the short-term and long-term effects of air pollution?

4. What are the three layers of soil?

5. What are three factors in the composition of soil?

PRACTICING SKILLS *(10 points each)* Study the diagram and answer the questions that follow.

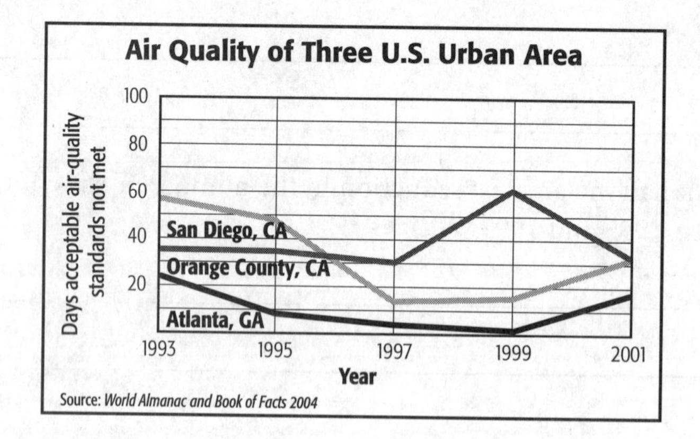

1. What was the net change in days of unacceptable air quality between 1995 and 1997 for San Diego?

2. Which location showed the least improvement in air quality from 1993 to 1999?

COMPOSING AN ESSAY *(30 points)* On a separate sheet of paper, write an essay in response to *one* of the following.

1. How have the locations, shapes, and sizes of landforms influenced human settlement and transportation?

2. Why might it be desirable to find alternatives to fossil fuel?

3. Describe the hydrologic cycle.

4. What is soil exhaustion and what can be done to prevent it?

Name ______________________ Class ______________ Date ____________

Chapter Test Form A

Human Geography

REVIEWING FACTS

MATCHING *(3 points each)* In the space provided, write the letter that matches each description. Choose your answers from the list below. Some answers will not be used.

______ **1.** When an idea spreads from one group to another

______ **2.** People who leave a country to live somewhere else

______ **3.** All the features of a people's way of life

______ **4.** Regional variety of a language

______ **5.** The statistical study of human populations

______ **6.** New ideas accepted by a culture

______ **7.** A language of trade and communication

______ **8.** When a group adopts some traits of another culture

______ **9.** The average number of people living in an area

______**10.** The belief in one god

a. demography
b. dialect
c. polytheism
d. diffusion
e. innovation
f. population density
g. monotheism
h. culture
i. emigrants
j. immigrants
k. acculturation
l. lingua franca

FILL IN THE BLANK *(3 points each)* For each of the following statements, fill in the blank with the appropriate word, phrase, or place.

1. The number of births each year for every 1,000 people living in a place is called the ______________________.

2. Southwest Asia and ______________________ have the highest rates of natural population increase.

3. A ______________________ is an area in which people have many shared culture traits.

4. ______________________ is the process of moving from one place to live in another.

5. English as the main language of Australia is an example of ______________________ diffusion.

6. Both Christianity and Islam are ______________________ religions.

7. The opposite of globalization is ______________________.

8. ______________________ are people who have been forced to leave their homes and cannot return.

9. When the ______________________ falls and the birthrate remains high, the total population grows.

10. The promise of free land to farmers in a different country is an example of a

______________________ factor.

UNDERSTANDING IDEAS *(3 points each)* For each of the following, place the letter of the best choice in the space provided.

______ **1.** Which of the following could be a push factor?
- **a.** country-wide famine
- **b.** good harvest
- **c.** increase in average income
- **d.** single ethnic group

______ **2.** Which of the following is not a common cause of culture change?
- **a.** trade
- **b.** homogeneity
- **c.** war
- **d.** migration

______ **3.** Individuals who help spread a religion are
- **a.** animists.
- **b.** acculturated.
- **c.** Jewish.
- **d.** missionaries.

______ **4.** Which of the following is not reflected in population density?
- **a.** size of population
- **b.** political climate of country
- **c.** size of country
- **d.** environmental conditions

______ **5.** The increase in connections between cultures around the world is called
- **a.** fundamentalism.
- **b.** ethnic diversity.
- **c.** globalization.
- **d.** acculturation.

PRACTICING SKILLS *(5 points each)* Study the diagram and answer the questions that follow.

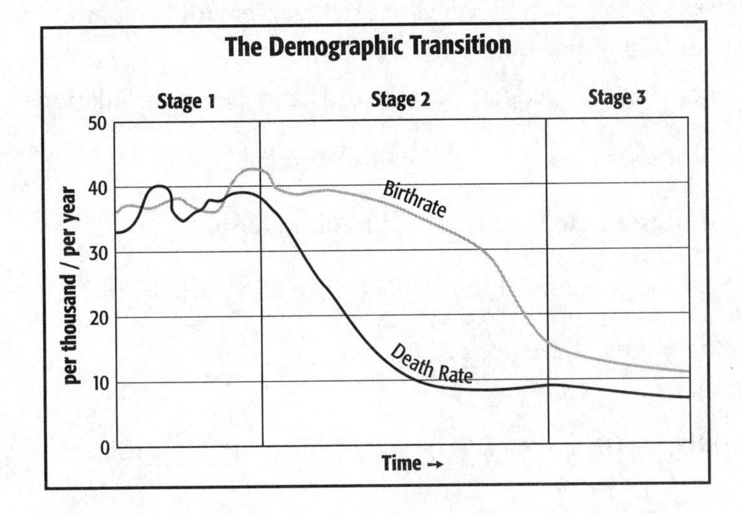

1. In which stage is the difference between the birthrate and the death rate the largest?

____

____

2. Why does the birthrate drop between Stages 2 and 3?

____

____

COMPOSING AN ESSAY *(15 points)* On a separate sheet of paper, write an essay in response to *one* of the following.

1. What factors have increased the rate of acculturation around the world?

2. Compare and contrast the population densities of North America and Southeast Asia.

Name ______________________ Class ____________ Date ____________

Chapter Test Form B

Human Geography

SHORT ANSWER *(10 points each)* Provide brief answers for each of the following. Use examples to support your answers.

1. List three factors used by geographers to study changes in populations.

2. Name three processes through which cultures change.

3. Compare and contrast globalization and traditionalism.

4. How has English become the lingua franca of much of the world?

5. Define the three types of religion.

PRACTICING SKILLS *(10 points each)* Study the diagram and answer the questions that follow.

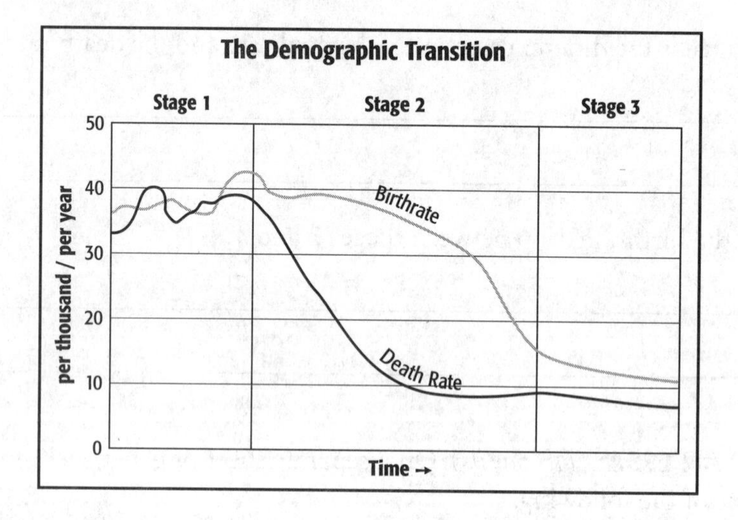

1. In which stage of demographic transition is the United States? Why?

2. Why are the death rate and birthrate so high in Stage 1?

COMPOSING AN ESSAY *(30 points)* On a separate sheet of paper, write an essay in response to *one* of the following.

1. Describe the model of demographic transition.

2. Why is the study of culture important?

3. Discuss the history of migration to the United States. Give examples of push and pull factors for the immigrants.

4. How do culture regions and political regions affect each other?

Name ______________________ Class ______________ Date ____________

Chapter Test Form A

Human Systems

REVIEWING FACTS

MATCHING *(3 points each)* In the space provided, write the letter that matches each description. Choose your answers from the list below. Some answers will not be used.

_____ **1.** The growth in the population of people living in towns and cities

_____ **2.** A system of roads, ports, and other facilities

_____ **3.** A feeling of pride and loyalty for one's country or culture group

_____ **4.** A system in which businesses, industries, and resources are privately owned

_____ **5.** The clearing of trees and brush for short-term farming

_____ **6.** Borders that are drawn with little regard to environmental or cultural patterns

_____ **7.** The growing of plants and raising of animals by humans for their own use

_____ **8.** Borders based on traits such as religion or language

_____ **9.** Taxes imposed on imports and exports

_____ **10.** A system in which the government owns or controls all the means of production

a. infrastructure
b. pastoralism
c. tariffs
d. shifting cultivation
e. domestication
f. communism
g. cultural boundaries
h. geometric boundaries
i. market economy
j. capitalism
k. urbanization
l. nationalism

FILL IN THE BLANK *(3 points each)* For each of the following statements, fill in the blank with the appropriate word, phrase, or place.

1. Free enterprise is the basis for ______________________.

2. People living in developed countries generally live in ______________________.

3. Roads, sewage treatment plants, and power lines are all parts of a country's

______________________.

4. A farm that grows large quantities of a single crop is an example of

______________________ agriculture.

5. ______________________ activities are not tied directly to resources, environmental conditions, or access to markets.

6. The borders between Ireland and Northern Ireland and between India and Pakistan are both ______________________ boundaries.

7. The most important international organization is the ______________________.

8. A country's ______________________ is measured by such factors as literacy rate, food consumption, and life expectancy.

9. Governments whose leaders are chosen through free elections are ______________________.

10. Pastoralism is one form of ______________________ agriculture.

UNDERSTANDING IDEAS *(3 points each)* For each of the following, write the letter of the best choice in the space provided.

______ **1.** The division of the cities of Minneapolis and St. Paul by the Mississippi River is an example of a(n)
- **a.** artificial boundary.
- **b.** geometric boundary.
- **c.** cultural boundary.
- **d.** natural boundary.

______ **2.** The most basic economic system is a
- **a.** subsistence economy.
- **b.** market economy.
- **c.** command economy.
- **d.** free enterprise economy.

______ **3.** Which country has a totalitarian government?
- **a.** Russia
- **b.** Great Britain
- **c.** Cuba
- **d.** Mexico

______ **4.** A steel manufacturing plant is an example of a
- **a.** tertiary activity.
- **b.** quaternary activity.
- **c.** primary activity.
- **d.** secondary activity.

______ **5.** The most important centers for economic power and wealth are
a. central business districts.
b. world cities.
c. edge cities.
d. agribusinesses.

PRACTICING SKILLS *(5 points each)* Study the table and answer the questions that follow.

Selected Countries' Statistics

		Per Capita GDP	Life Expectancy	Literacy Rate	Urban	TV Sets (per 1,000 persons)	Physicians
Developed Countries	Australia	$21,200	77, male; 83, female	100%	85%	639	1 per 389 persons
	Japan	$23,100	77, male; 83, female	100%	79%	708	1 per 522 persons
	United States	$31,500	73, male; 80, female	97%	76%	847	1 per 365 persons
Developing Countries	Afghanistan	$ 800	48, male; 47, female	31.5%	21%	10	1 per 6,690 persons
	Haiti	$ 1,300	50, male; 54, female	45%	35%	4	1 per 9,846 persons
	Mali	$ 790	47, male; 49, female	31%	29%	12	1 per 18,376 persons
Middle-Income Countries	Brazil	$ 6,100	59, male; 68, female	85%	81%	317	1 per 681 persons
	Mexico	$ 8,300	69, male; 76, female	90%	74%	257	1 per 613 persons
	Thailand	$ 6,100	66, male; 73, female	94%	21%	54	1 per 3,461 persons

Source: *The World Almanac and Book of Facts 2000, 2001*

1. What relationship exists between the level of people living in urban areas and the number of television sets per 1,000 people? Why?

____

____

2. What country has the lowest ratio of physicians to people?

____

____

COMPOSING AN ESSAY *(15 points)* On a separate sheet of paper, write an essay in response to *one* of the following.

1. How do the three types of economic systems relate to the economic development of countries?

2. How do urban and rural geography affect each other?

Name ______________________ Class ____________ Date ____________

Chapter Test Form B

Human Systems

SHORT ANSWER *(10 points each)* Provide brief answers for each of the following. Use examples to support your answers.

1. List the three main types of economic systems.
2. Describe four ways a country's level of development can be measured.
3. Describe how cities have changed over the past 200 years.
4. List three factors used to select the location of a city.
5. List three types of geographic boundaries and give an example of each.

PRACTICING SKILLS *(10 points each)* Study the table and answer the questions that follow.

Selected Countries' Statistics

		Per Capita GDP	Life Expectancy	Literacy Rate	Urban	TV Sets (per 1,000 persons)	Physicians
Developed Countries	Australia	$21,200	77, male; 83, female	100%	85%	639	1 per 389 persons
	Japan	$23,100	77, male; 83, female	100%	79%	708	1 per 522 persons
	United States	$31,500	73, male; 80, female	97%	76%	847	1 per 365 persons
Developing Countries	Afghanistan	$ 800	48, male; 47, female	31.5%	21%	10	1 per 6,690 persons
	Haiti	$ 1,300	50, male; 54, female	45%	35%	4	1 per 9,846 persons
	Mali	$ 790	47, male; 49, female	31%	29%	12	1 per 18,376 persons
Middle-Income Countries	Brazil	$ 6,100	59, male; 68, female	85%	81%	317	1 per 681 persons
	Mexico	$ 8,300	69, male; 76, female	90%	74%	257	1 per 613 persons
	Thailand	$ 6,100	66, male; 73, female	94%	21%	54	1 per 3,461 persons

Source: *The World Almanac and Book of Facts 2000, 2001*

1. Compare and contrast the life expectancies of people who live in developed and developing countries.
2. How is a country's literacy rate tied to its gross domestic product?

COMPOSING AN ESSAY *(30 points)* On a separate sheet of paper, write an essay in response to *one* of the following.

1. How are education rates and standards of living related to economic activities?
2. How are a country's governmental stability and economic system related?
3. What factors allow a country to move out of the developing country classification?
4. How might the drawing of geographic boundaries affect a country's relationships with other countries?

Name ______________________ Class ______________ Date ____________

Unit Test Form A

The Geographer's World

REVIEWING FACTS

MATCHING *(3 points each)* In the space provided, write the letter that matches each description. Choose your answers from the list below. Some answers will not be used.

______ **1.** The physical, human, and cultural features of a place

______ **2.** The solid crust of Earth

______ **3.** The way a person looks at something

______ **4.** The meeting of two air masses of widely different temperature or moisture

______ **5.** An area covered by a river and its tributaries

______ **6.** The breaking down and decay of rocks over time

______ **7.** The statistical study of human populations

______ **8.** A system in which businesses, industries, and resources are privately owned

______ **9.** All the features of a people's way of life

______ **10.** The weather conditions in a geographic area over a long period of time

a. perspective
b. equinox
c. front
d. watershed
e. rotation
f. culture
g. capitalism
h. demography
i. lithosphere
j. landscape
k. climate
l. weathering

FILL IN THE BLANK *(3 points each)* For each of the following statements, fill in the blank with the appropriate word, phrase, or place.

1. A country's ______________________ is measured by such factors as literacy rate, food consumption, and life expectancy.

2. The two main branches of geography are human geography and ______________________ geography.

3. You would expect to experience the most precipitation in persistent ______________________-pressure zones.

4. ______________________ is the process of moving from one place to live in another.

5. The theory of ______________________ explains how forces within the planet create landforms.

6. The shape of Earth's orbit around the Sun is ______________________.

7. Contour plowing is one way to reduce soil ______________________.

8. Earth's ______________________ holds the atmosphere around the planet.

9. When the ______________________ falls and the birthrate remains high, the total population increases.

10. Ocean currents are caused by wind, ______________________, and varying ocean temperatures.

UNDERSTANDING IDEAS *(3 points each)* For each of the following, write the letter of the best choice in the space provided.

______ **1.** Geographers use what type of perspective to study the world?
- **a.** spatial
- **b.** linear
- **c.** intuitive
- **d.** digital

______ **2.** A steel manufacturing plant is an example of what type of activity?
- **a.** primary
- **b.** secondary
- **c.** tertiary
- **d.** quaternary

______ **3.** The increase in connections between cultures around the world is called
- **a.** fundamentalism.
- **b.** ethnic diversity.
- **c.** acculturation.
- **d.** globalization.

______ **4.** Which of the following is the most basic type of economic system?
- **a.** subsistence
- **b.** market
- **c.** command
- **d.** free enterprise

Name ______________________ Class ______________ Date ______________

______ **5.** What affects the amount of solar energy that difference places on Earth receive?
- **a.** Earth's revolution
- **b.** time zones
- **c.** tilt of Earth's axis
- **d.** Earth's rotation

PRACTICING SKILLS *(5 points each)* Study the contour map and answer the questions that follow.

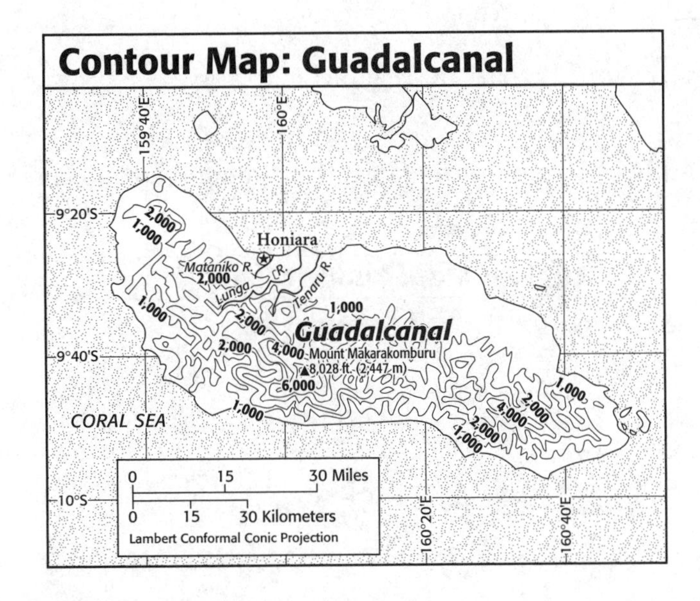

1. What is the approximate longitude and latitude of the capital city of Guadalcanal?

____

____

2. What is the elevation in the area of the Lunga River?

____

____

COMPOSING AN ESSAY *(10 points)* On a separate sheet of paper, write an essay in response to *one* of the following.

1. Describe the six essential elements of geography and how they relate to the five themes of geography.

2. What factors have increased the rate of acculturation around the world?

Name ______________________ Class ______________ Date ______________

Unit Test Form B

The Geographer's World

SHORT ANSWER *(10 points each)* Provide brief answers for each of the following. Use examples to support your answers.

1. List the four parts of the Earth system.

2. What are three types of plate movement?

3. List three factors used by geographers to study changes in population.

4. Describe the three types of economic systems.

5. Describe how geographers have made it possible to accurately locate places on Earth.

PRACTICING SKILLS *(10 points each)* Study the contour map and answer the questions that follow.

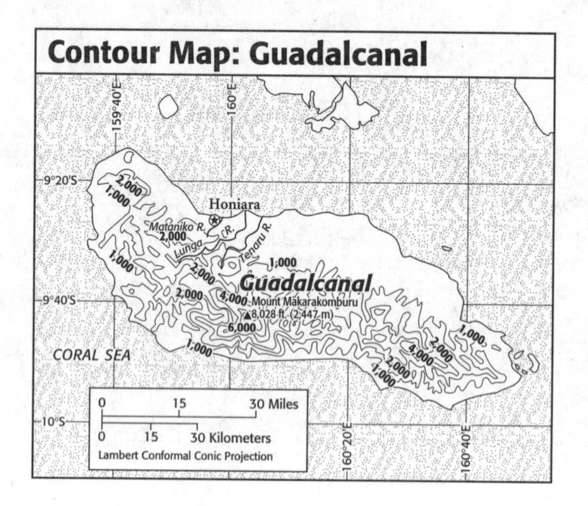

1. Based on this map, describe Guadalcanal.

2. Where is the sharpest change in elevation?

COMPOSING AN ESSAY *(30 points)* On a separate sheet of paper, write an essay in response to *one* of the following.

1. What key factors must be achieved by a country for it to rise out of developing country status?

2. Compare and contrast the tropical humid climate and the tropical wet and dry climate.

3. Describe the three types of map projections and how they are different.

4. How have the locations, shapes, and sizes of landforms influenced human settlement and transportation?

Name ______________________ Class ______________ Date ____________

Chapter Test Form A

Natural Environments of North America

REVIEWING FACTS

MATCHING *(3 points each)* In the space provided, write the letter of the term or place that matches each description. Some answers will not be used.

______ **1.** Area at or near the foot of a mountain range

______ **2.** Lower area of land surrounded by mountains

______ **3.** Point where rivers flow from hard rock to soft rock of coastal plain

______ **4.** Region of the United States that lies completely within the tropics

______ **5.** Extremely fertile soil deposited by rivers or streams

______ **6.** Events in the physical environment that can destroy human life and property

______ **7.** A place where magma wells up to the surface from Earth's mantle

______ **8.** River used for both irrigation and to produce electricity

______ **9.** Small plants that consist of algae and fungi

______ **10.** One of the most famous fishing areas in the world

a. fall line
b. alluvial soil
c. Colorado River
d. basin
e. Grand Banks
f. hot spot
g. Hawaiian Islands
h. Piedmont
i. St. Lawrence River
j. natural hazards
k. Mississippi River
l. lichens

FILL IN THE BLANK *(3 points each)* For each of the following statements, fill in the blank with the appropriate word, phrase, or place.

1. Areas east of the Sierra Nevada and Cascades are dry due to the

______________________.

2. Most major landforms in North America stretch from ______________________.

3. North America's major river systems flow east or west of the

______________________.

4. Canada's major oil-producing regions are located in ______________________.

5. Only the ______________________ climate is not found in the United States.

6. Plant and animal life are greatly influenced by ______________________ patterns.

7. Canadian forests provide lumber, pulpwood, and ______________________.

8. The main commercial fish on the Pacific coast is ______________________.

9. Major copper deposits are found in ______________________.

10. The Hawaiian Islands are located on a ______________________.

UNDERSTANDING IDEAS *(3 points each)* For each of the following, write the letter of the best choice in the space provided.

______ 1. Which of the following is not a volcano in the Cascades?
- **a.** Mount McKinley
- **b.** Mount Saint Helens
- **c.** Mount Hood
- **d.** Mount Shasta

______ 2. Which region is not rich in alluvial soil?
- **a.** Mississippi Valley
- **b.** Imperial Valley
- **c.** Rio Grande Valley
- **d.** Mesabi Range

______ 3. What type of climate is found in most of the southeastern quarter of the United States?
- **a.** ice cap
- **b.** tropical wet and dry
- **c.** humid subtropical
- **d.** humid continental

______ 4. Which city is located on the fall line?
- **a.** Philadelphia, Pennsylvania
- **b.** Los Angeles, California
- **c.** New York City
- **d.** Seattle, Washington

______ 5. What type of vegetation is most commonly found in humid areas?
- **a.** scrub
- **b.** grasslands
- **c.** forests
- **d.** desert

PRACTICING SKILLS *(5 points each)* Study the map and answer the questions that follow.

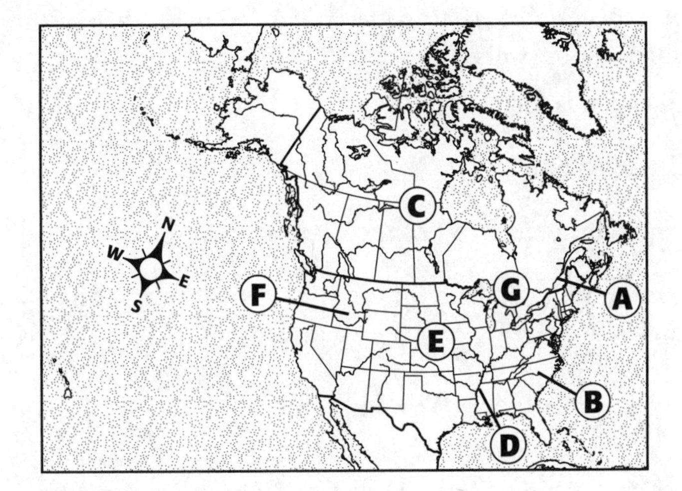

1. What letter marks the Canadian Shield?

____

2. What letter marks the Rocky Mountains?

____

COMPOSING AN ESSAY *(15 points)* On a separate sheet of paper, write an essay in response to *one* of the following.

1. How did the fall line influence early colonial settlements?

2. How have the natural resources of North America contributed to the United States and Canada's statuses as developed countries?

Name ______________________ Class ______________ Date ______________

Chapter Test Form B

Natural Environments of North America

SHORT ANSWER *(10 points each)* Provide brief answers for each of the following. Use examples to support your answers.

1. How are barrier islands formed?

2. What is the Canadian Shield?

3. Name four mountain ranges west of the Great Plains.

4. Why does the intermountain area on the West Coast of the United States have a variety of climates?

5. What are the two main climates found in Canada?

PRACTICING SKILLS *(10 points each)* Study the table and answer the questions that follow.

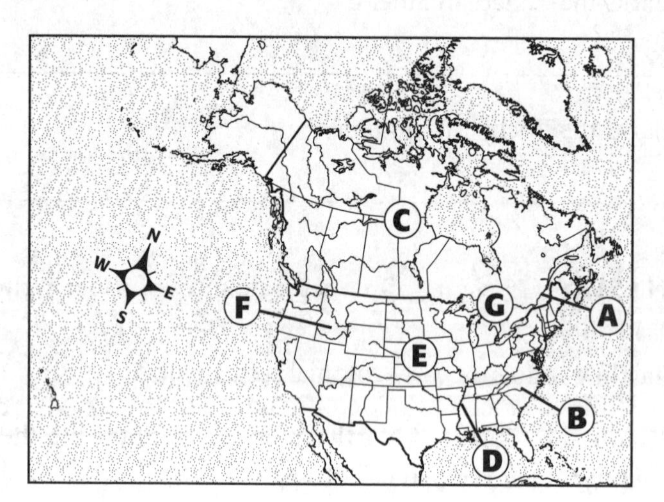

1. What letters mark rivers?

2. What letter represents the Piedmont?

COMPOSING AN ESSAY *(30 points)* On a separate sheet of paper, write an essay in response to *one* of the following.

1. Describe the major river systems of North America.

2. How do you think the location of mountain ranges in the United States has affected development and settlement?

3. How does the Gulf Stream affect the climate of the southeastern United States?

4. Compare and contrast the landforms east and west of the Great Plains.

Name ______________________ Class ______________ Date ______________

Chapter Test Form A

The United States

REVIEWING FACTS

MATCHING *(3 points each)* In the space provided, write the letter of the term or place that matches each description. Some answers will not be used.

_____ **1.** Restoration of rundown urban residential areas

_____ **2.** Large farms that produce one major crop

_____ **3.** Group of cities that have grown in to one large, built-up area

_____ **4.** Territory controlled by people from a foreign land

_____ **5.** Large powerful country

_____ **6.** Able to speak two languages

_____ **7.** Land fit for growing crops

_____ **8.** When value of a country's imports exceeds the value of its exports

_____ **9.** Third-largest city in the United States

_____**10.** Cloth products

a. bilingual
b. textiles
c. trade deficit
d. Chicago
e. plantations
f. NAFTA
g. Seattle
h. colonies
i. Megalopolis
j. gentrification
k. arable
l. superpower

FILL IN THE BLANK *(3 points each)* For each of the following statements, fill in the blank with the appropriate word, phrase, or place.

1. Most people in the United States who are bilingual speak English and ______________________.

2. Western cities strictly enforce building codes to limit the damage from ______________________.

3. Most new immigrants to the United States now come from ______________________ and ______________________.

4. The earliest settlers of North America came across a land bridge that linked Alaska and ______________________.

5. The largest metropolitan area in the South is ______________________.

6. The ______________________ and Dairy Belts are located in the Midwest.

7. Regional economies in the ______________________ colonies were based on cotton and tobacco.

8. The use of ______________________ by farmers has caused excessive algae growth in the Gulf of Mexico.

9. The ______________________ is the most industrialized region of the United States.

10. The ______________________ is home to the largest percentage of the population in the United States.

UNDERSTANDING IDEAS *(3 points each)* For each of the following, write the letter of the best choice in the space provided.

______ **1.** What has happened to trade in North America since the signing of NAFTA?
- **a.** It has stayed the same.
- **b.** It has increased dramatically.
- **c.** It has decreased dramatically.
- **d.** It has created large trade deficits.

______ **2.** Which of the following is not a type of music that originated in the United States?
- **a.** classical
- **b.** jazz
- **c.** rap
- **d.** country

______ **3.** Most early colonial settlements were located
- **a.** in the South.
- **b.** on natural harbors or navigable rivers.
- **c.** west of the Appalachians.
- **d.** far from Indian Territory.

______ **4.** What do some people consider a negative effect of gentrification?
- **a.** increased traffic congestion
- **b.** reduced commuting times
- **c.** increased property taxes in poor urban areas
- **d.** decreased property taxes in suburban areas

______ **5.** What city has the country's most important railroad hub?
- **a.** St. Louis
- **b.** Chicago
- **c.** Detroit
- **d.** Philadelphia

Name ______________________ Class ______________ Date ____________

PRACTICING SKILLS *(5 points each)* Study the graph and answer the questions that follow.

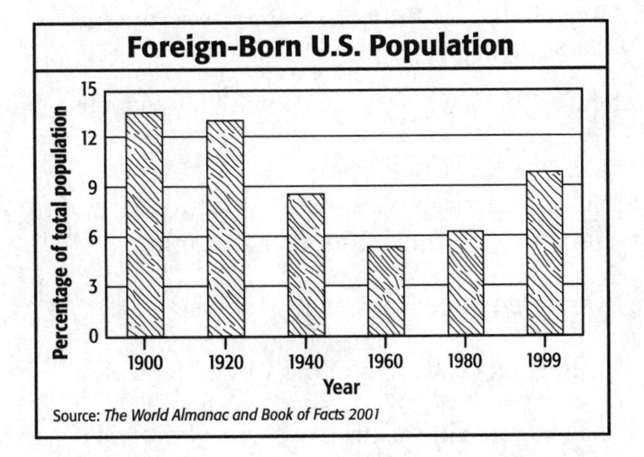

1. In what year was the percentage of the U.S. population that was foreign-born highest?

____

2. What events may have affected the percentage of foreign-born citizens in the United States in 1940?

____

COMPOSING AN ESSAY *(15 points)* On a separate sheet of paper, write an essay in response to *one* of the following.

1. How did the unique development of cultural patterns in the United States lead to innovations in music, architecture, and literature?

2. How do settlement patterns reflect land use in the United States?

Name ______________________ Class ______________ Date ____________

Chapter Test Form B

The United States

SHORT ANSWER *(10 points each)* Provide brief answers for each of the following. Use examples to support your answers.

1. Why did the leaders of the newly formed United States create a federal style of government?

2. Give three examples of American place-names that are derived from Indian names, three from British names, and three from French names.

3. What states are considered to be part of the Midwest?

4. Name three environmental challenges in the United States.

5. What were four major events in the history of the United States that took place in the 1900s?

PRACTICING SKILLS *(10 points each)* Study the graph and answer the questions that follow.

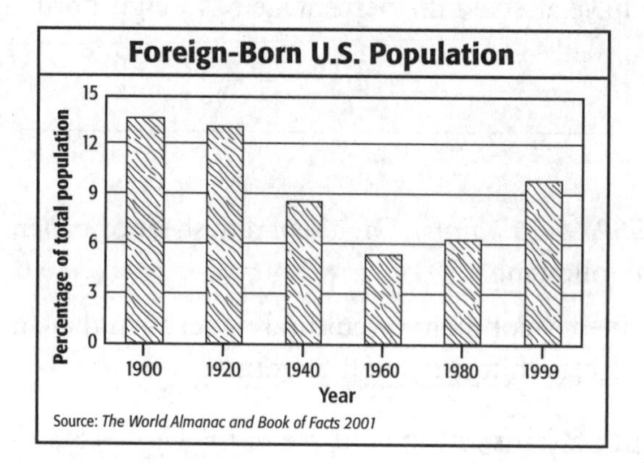

1. In which year was the smallest percentage of the U.S. population foreign-born?

2. What would you predict the graph to look like in 2010?

COMPOSING AN ESSAY *(30 points)* On a separate sheet of paper, write an essay in response to *one* of the following.

1. What geographic, cultural, and physical characteristics have allowed the United States to become a superpower?

2. How has the ethnic and cultural composition of the United States changed in the last century?

3. Why have many new industries moved into the South in recent years?

4. Describe the early economies of the colonies.

Name ______________________ Class ______________ Date ____________

Chapter Test Form A

Canada

REVIEWING FACTS

MATCHING *(3 points each)* In the space provided, write the letter of the term or place that matches each description. Some answers will not be used.

_____ **1.** Canadian governmental district

_____ **2.** Ontario's capital

_____ **3.** Important oil and agricultural city in Alberta

_____ **4.** The capital of Canada

_____ **5.** General agreement

_____ **6.** Belief that parts of the country should be independent

_____ **7.** Region far from major population centers

_____ **8.** Canada's elected legislature

_____ **9.** Manitoba's capital and chief city of Prairie Provinces

_____ **10.** Strong loyalty to one's own region

a. province
b. consensus
c. Winnipeg
d. Vancouver
e. regionalism
f. hinterland
g. Ottawa
h. separatism
i. Toronto
j. Windsor
k. Calgary
l. parliament

FILL IN THE BLANK *(3 points each)* For each of the following statements, fill in the blank with the appropriate word, phrase, or place.

1. The economy of Canada is based on ______________________ and service industries.

2. The economic relationship between Canada and the United States has strengthened since the signing of ______________________.

3. Canada's poorest region is the ______________________ provinces.

4. If Quebec were to attain independence, the ______________________ provinces would be cut off from the rest of the country.

5. The ______________________ were the first Europeans to explore Canada.

6. Canadians' strong sense of ______________________ is partly due to physical geography.

7. The most recently created territory is ______________________.

8. Canada has a ______________________ economy.

9. ______________________ is Canada's main Pacific port.

10. ______________________ was the first European explorer of Canada's interior.

UNDERSTANDING IDEAS *(3 points each)* For each of the following, write the letter of the best choice in the space provided.

_____ **1.** The British government created the self-governing dominion of Canada in
- **a.** 1763.
- **b.** 1806.
- **c.** 1867.
- **d.** 1879.

_____ **2.** What are the main natural resources of the Atlantic Provinces?
- **a.** fishing and tourism
- **b.** mineral deposits and oil
- **c.** mining and good soil
- **d.** timber and fishing

_____ **3.** In what province are the Rocky Mountains located?
- **a.** Alberta
- **b.** British Columbia
- **c.** Canadian North
- **d.** Saskatchewan

_____ **4.** Who is the most important buyer of British Columbia's forest products?
- **a.** Great Britain
- **b.** the United States
- **c.** Japan
- **d.** Quebec and Ontario

_____ **5.** Which city is a major port on the St. Lawrence Seaway?
- **a.** Windsor
- **b.** Montreal
- **c.** Ottawa
- **d.** Toronto

PRACTICING SKILLS *(5 points each)* Study the table and answer the questions that follow.

Hectares of Forest Area Harvested in Canada (in thousands)			
Province	**1993**	**1994**	**1995**
British Columbia	210	190	190
Manitoba	9	10	12
Northwest Territories	2	2	2
Ontario	219	215	218

1. Which province had the most consistent harvest over time?

____

2. What factors might contribute to the Northwest Territories' small harvest?

____

COMPOSING AN ESSAY *(15 points)* On a separate sheet of paper, write an essay in response to one of the following.

1. Take a position and argue whether you think Quebec should be allowed to become independent from Canada.

2. Describe Canada, taking the viewpoints of an Inuit, a French Canadian, and an English descendant.

Name ______________________ Class ______________ Date ____________

Chapter Test Form B

Canada

SHORT ANSWER *(10 points each)* Provide brief answers for each of the following. Use examples to support your answers.

1. What European countries were most influential in shaping Canada's development?
2. Why are Quebec and Ontario considered the heartland of Canada?
3. What has contributed to the growth of the major cities in the Prairie Provinces?
4. What is Nunavut, and how and why was it created?
5. Explain how Quebec is culturally different from the rest of Canada.

PRACTICING SKILLS *(10 points each)* Study the table and answer the questions that follow.

Hectares of Forest Area Harvested in Canada (in thousands)

Province	1993	1994	1995
British Columbia	210	190	190
Manitoba	9	10	12
Northwest Territories	2	2	2
Ontario	219	215	218

1. Which area had the least forest area harvested in 1995?
2. Which province had the largest decrease in forest area harvested between 1993 and 1995?

COMPOSING AN ESSAY *(30 points)* On a separate sheet of paper, write an essay in response to one of the following.

1. What does the United States have in common with Canada?
2. What is the most densely populated region of Canada and why is this so?
3. How would Canada be affected if Quebec won its independence?
4. Why can it be said that life in the Atlantic Provinces is challenging?

Name ______________________ Class ______________ Date ____________

Unit Test Form A

The United States and Canada

REVIEWING FACTS

MATCHING *(3 points each)* In the space provided, write the letter of the term or place that matches each description. Some answers will not be used.

______ **1.** Third-largest city in the United States

______ **2.** A place where magma wells up to the surface from Earth's mantle

______ **3.** The capital of Canada

______ **4.** Restoration of rundown urban residential areas

______ **5.** Belief that parts of the country should be independent

______ **6.** River used for both irrigation and to produce electricity

______ **7.** When value of a country's imports exceeds the value of its exports

______ **8.** Area at or near the foot of a mountain range

______ **9.** General agreement

______ **10.** Point where rivers flow from hard rock to soft rock of coastal plain

a. Chicago
b. Megalopolis
c. fall line
d. consensus
e. separatism
f. hot spot
g. trade deficit
h. Piedmont
i. Ottawa
j. regionalism
k. gentrification
l. Colorado River

FILL IN THE BLANK *(3 points each)* For each of the following statements, fill in the blank with the appropriate word, phrase, or place.

1. The economic relationship between Canada and the United States has strengthened since the signing of ______________________.

2. The ______________________ is home to the largest percentage of the population in the United States.

3. Areas east of the Sierra Nevada and Cascades are dry due to the ______________________.

4. Most new immigrants to the United States now come from ______________________ and ______________________.

5. Canada's strong sense of ______________________ is partly due to physical geography.

6. Major copper deposits are found in ______________________.

7. The ______________________ and Dairy Belts are located in the Midwest.

8. North America's major river systems flow east or west of the ______________________.

9. The most recently created territory in Canada is ______________________.

10. ______________________ is Canada's main Pacific port.

UNDERSTANDING IDEAS *(3 points each)* For each of the following, write the letter of the best choice in the space provided.

______ **1.** What type of climate is found in most of the southeastern quarter of the United States?
- **a.** ice cap
- **b.** tropical wet and dry
- **c.** humid subtropical
- **d.** humid continental

______ **2.** What do some people consider a negative effect of gentrification?
- **a.** increased traffic congestion
- **b.** reduced commuting times
- **c.** increased property taxes in poor urban areas
- **d.** decreased property taxes in suburban areas

______ **3.** Who is the most important buyer of British Columbia's forest products?
- **a.** Great Britain
- **b.** the United States
- **c.** Japan
- **d.** Quebec and Ontario

______ **4.** What has happened to trade in North America since the signing of NAFTA?
- **a.** It has stayed the same.
- **b.** It has increased dramatically.
- **c.** It has decreased dramatically.
- **d.** It has created large trade deficits.

______ **5.** Which region is not rich in alluvial soil?
a. Mississippi Valley
b. Imperial Valley
c. Rio Grande Valley
d. Mesabi Range

PRACTICING SKILLS *(5 points each)* Study the table and answer the questions that follow.

Hectares of Forest Area Harvested in Canada (in thousands)

Province	1993	1994	1995
British Columbia	210	190	190
Manitoba	9	10	12
Northwest Territories	2	2	2
Ontario	219	215	218

1. Which province had the most consistent harvest over time?

____

____

2. What factors might contribute to the Northwest Territories' small harvest?

____

____

COMPOSING AN ESSAY *(15 points)* On a separate sheet of paper, write an essay in response to *one* of the following.

1. Describe Canada, taking the viewpoints of an Inuit, a French Canadian, and an English descendant.

2. How did the unique development of cultural patterns in the United States lead to innovations in music, architecture, and literature?

Name ______________________ Class ______________ Date ______________

Unit Test Form B

The United States and Canada

SHORT ANSWER *(10 points each)* Provide brief answers for each of the following. Use examples to support your answers.

1. What is Nunavut, and how and why was it created?
2. Name four mountain ranges west of the Great Plains.
3. Why did the leaders of the newly formed United States create a federal style of government?
4. What has contributed to the growth of the major cities in the Prairie Provinces?
5. Name three environmental challenges in the United States.

PRACTICING SKILLS *(10 points each)* Study the table and answer the questions that follow.

Hectares of Forest Area Harvested in Canada (in thousands)

Province	1993	1994	1995
British Columbia	210	190	190
Manitoba	9	10	12
Northwest Territories	2	2	2
Ontario	219	215	218

1. Which province had the least forest area harvested in 1995?
2. Which province had the largest decrease in forest area harvested?

COMPOSING AN ESSAY *(30 points)* On a separate sheet of paper, write an essay in response to *one* of the following.

1. What geographic, cultural, and physical characteristics have allowed the United States to become a superpower?
2. How would Canada be affected if Quebec won its independence?
3. Compare and contrast the landforms east and west of the Great Plains.
4. What does the United States have in common with Canada?

Name ______________________ Class ______________ Date ____________

Chapter Test Form A

Mexico

REVIEWING FACTS

MATCHING *(3 points each)* In the space provided, write the letter that matches each description. Choose your answers from the list below. Some answers will not be used.

_____ **1.** A large estate worked by peasants

_____ **2.** Factories in Mexico that produce goods for U.S. markets

_____ **3.** People of mixed European and Indian heritage

_____ **4.** A narrow strip of land that connects two larger land areas

_____ **5.** The highest mountain in Mexico

_____ **6.** Crops grown for sale in a market

_____ **7.** A large estate farmed by workers who live on the property

_____ **8.** The cultural, political, and economic center of Mexico

_____ **9.** An important seaport and communications center

_____ **10.** A steep-sided depression that forms when a cave roof collapses

a. isthmus
b. Veracruz
c. Orizaba
d. conquistadores
e. mestizos
f. plantation
g. hacienda
h. Mexico City
i. cash crops
j. sinkhole
k. Guadalajara
l. maquiladoras

FILL IN THE BLANK *(3 points each)* For each of the following statements, fill in the blank with the appropriate word, phrase, or place.

1. Mexico's poor ______________________ is a major obstacle to continued economic growth.

2. One result of the Mexican Revolution was ______________________ reform.

3. ______________________ is a style of cultivation in which several crops are grown together in a single plot.

4. The Yucatán Peninsula lies atop ______________________, which makes the area prone to sinkholes.

5. The ______________________ separates the Gulf of Mexico and the Pacific Ocean.

6. The ______________________ mountains run along the southern Pacific coast.

7. ______________________ are Mexicans of mixed European and Indian ancestry.

8. Mexico City has serious air pollution problems that are caused in part by the ring of ______________________ that surround the city.

9. Spanish soldiers brought many diseases, particularly ______________________, to the indigenous population of Mexico.

10. ______________________ is the poorest region in Mexico.

UNDERSTANDING IDEAS *(3 points each)* For each of the following, write the letter of the best choice in the space provided.

_____ **1.** Which region of Mexico contains deposits of oil and natural gas?
- **a.** central Mexico
- **b.** Gulf lowlands
- **c.** southern Mexico
- **d.** northern Mexico

_____ **2.** What does NAFTA stand for?
- **a.** North American Free Trade Agreement
- **b.** North American Food and Transportation Authority
- **c.** North American Foreign Travel Association
- **d.** North American Federal Transportation Agreement

_____ **3.** When did the Mexican Revolution take place?
- **a.** 1910–20
- **b.** 1810–20
- **c.** 1920–30
- **d.** 1901–07

_____ **4.** What was the capital city of the Aztecs?
- **a.** Mazatlán
- **b.** Tampico
- **c.** Campeche
- **d.** Tenochtitlán

_____ **5.** What is the largest landform in Mexico?
- **a.** Isthmus of Tehuantepec
- **b.** Sierra Madre del Sur
- **c.** Mt. Orizaba
- **d.** Mexican Plateau

PRACTICING SKILLS *(5 points each)* Study the map and answer the questions that follow.

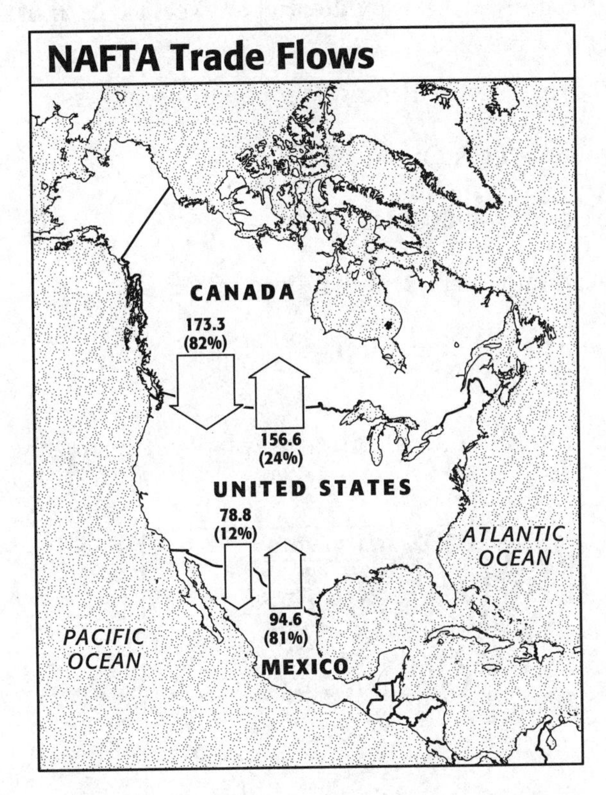

1. Which country exports the largest percent of its total exports to the United States?

____

____

2. With which country does the United States trade more?

____

____

COMPOSING AN ESSAY *(15 points)* On a separate sheet of paper, write an essay in response to *one* of the following.

1. Explain why Spanish conquistadores found it so easy to conquer Mexico.

2. Discuss how daily life in Mexico has changed with economic growth.

Name ______________________ Class ______________ Date ____________

Chapter Test Form B

Mexico

SHORT ANSWER *(10 points each)* Provide brief answers for each of the following. Use examples to support your answers.

1. What factors make the Yucatán Peninsula prone to sinkholes?

2. What three mountain ranges ring Mexico City?

3. What are Mexico's leading industries?

4. Name five American Indian cultures that lived in Mexico.

5. How was the *ejido* system different from the plantation system?

PRACTICING SKILLS *(10 points each)* Study the map and answer the questions that follow.

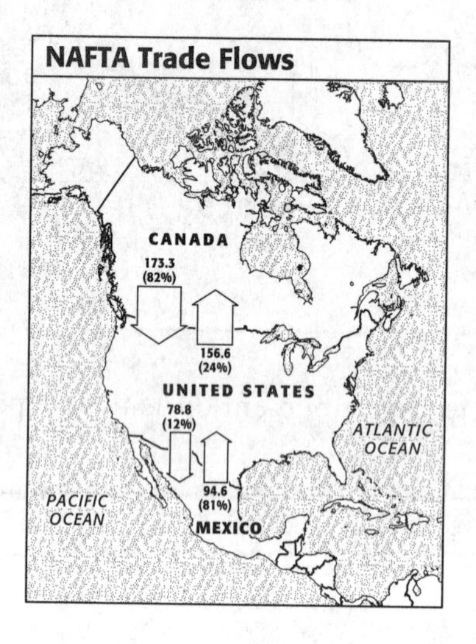

1. Which country has a larger difference between its exports to and imports from the United States?

2. What is the difference in U.S. dollars between Mexico's imports and exports?

COMPOSING AN ESSAY *(30 points)* On a separate sheet of paper, write an essay in response to *one* of the following.

1. Describe how a typical Mexican colonial town evolved and was organized.

2. Discuss why the majority of Mexicans today are mestizos.

3. Describe the important challenges facing Mexico today.

4. Discuss the landforms that influence Mexico's climate. Why is there such a broad range of climates in the country?

Name ______________________ Class ______________ Date ____________

Chapter Test Form A

Central America and the Caribbean

REVIEWING FACTS

MATCHING *(3 points each)* In the space provided, write the letter of the term or place that matches each description. Some answers will not be used.

_____ **1.** A term that means "native"

_____ **2.** Island containing Haiti and the Dominican Republic

_____ **3.** People of African and European descent

_____ **4.** Caribbean Community and Common Market

_____ **5.** Form of voodoo that originated in Cuba

_____ **6.** Ore from which aluminum is made

_____ **7.** Blended version of African religious beliefs and Christianity

_____ **8.** Blend of European and African or Caribbean Indian languages

_____ **9.** Tree from which cocoa beans are harvested

_____ **10.** Tree whose roots grow in salt water

a. indigenous
b. mangrove
c. CARICOM
d. creole
e. Hispaniola
f. bauxite
g. ecotourism
h. cacao
i. voodoo
j. mulattos
k. Jamaica
l. Santeria

FILL IN THE BLANK *(3 points each)* For each of the following statements, fill in the blank with the appropriate word, phrase, or place.

1. Cuba has historically had very close ties to ______________________.

2. ______________________ is Central America's least developed country.

3. Europeans established ______________________ plantations in the Caribbean.

4. One of Central America's greatest natural resources is ______________________, the basis for its tourism industry.

5. The largest African populations in the Caribbean are found on

______________________ and ______________________.

6. ______________________ is the only non-Spanish speaking country in Central America.

7. ______________________ has the region's highest standard of living.

8. Most of the region's Central American Indians now live in

______________________.

9. The Panama Canal was controlled by the ______________________ until 1999.

10. The landforms of Central America have been shaped by ______________________.

UNDERSTANDING IDEAS *(3 points each)* For each of the following, write the letter of the best choice in the space provided.

_____ **1.** The United States took what two islands from Spain in 1898?
a. Cuba and Hispaniola
b. Jamaica and the Dominican Republic
c. Puerto Rico and Cuba
d. Puerto Rico and Haiti

_____ **2.** Which island was formed from volcanoes?
a. Martinique
b. Bahamas
c. Hispaniola
d. Cuba

_____ **3.** What is the most industrialized country in the Caribbean?
a. Cuba
b. Dominican Republic
c. Puerto Rico
d. Bahamas

_____ **4.** Which of the following is not one of the tectonic plates in this region?
a. Cocos Plate
b. North African Plate
c. North American Plate
d. Caribbean Plate

_____ **5.** Which of the following is not an important commercial crop in the region?
a. cacao
b. sugar
c. coffee
d. rice

Name ______________________ Class ______________ Date ______________

PRACTICING SKILLS *(5 points each)* Study the diagram and answer the questions that follow.

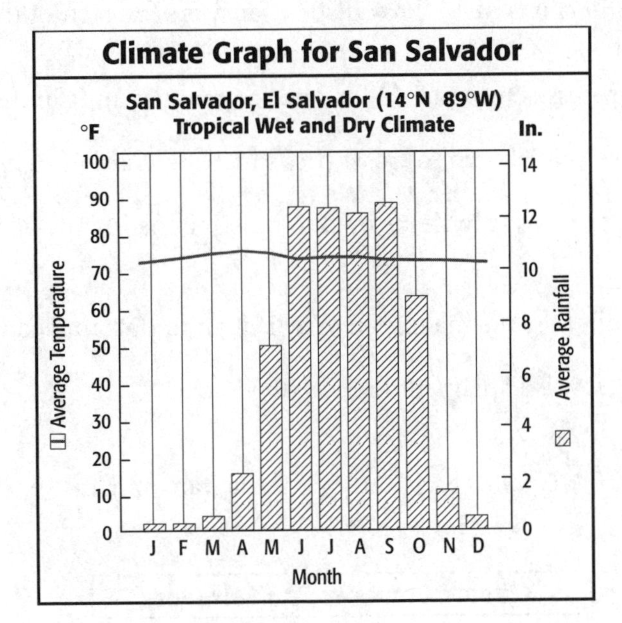

1. About how much total rainfall does San Salvador receive in the spring months?

____

____

2. In what month does San Salvador receive the most rain?

____

____

COMPOSING AN ESSAY *(15 points)* On a separate sheet of paper, write an essay in response to *one* of the following.

1. Compare and contrast Cuba's economy with the economies of the other Caribbean countries.

2. What are the benefits of promoting ecotourism as an important part of Costa Rica's economy?

Name ______________________ Class ______________ Date ______________

Chapter Test Form B

Central America and the Caribbean

SHORT ANSWER *(10 points each)* Provide brief answers for each of the following. Use examples to support your answers.

1. What physical process is responsible for creating many islands in the Lesser Antilles?

2. What factors have contributed to Costa Rica's high standard of living?

3. Explain why the term West Indies is used to describe the islands of the Caribbean region.

4. What country built the Panama Canal? Why is it an important economic resource?

5. What is Creole? How is Papiamento related to Creole?

PRACTICING SKILLS *(10 points each)* Study the diagram and answer the questions that follow.

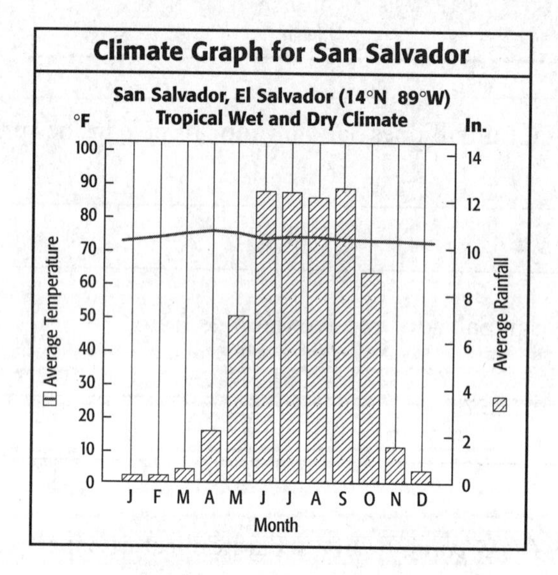

1. In what months does San Salvador receive the least amount of rain?

2. What is the approximate annual rainfall of San Salvador?

COMPOSING AN ESSAY *(30 points)* On a separate sheet of paper, write an essay in response to *one* of the following.

1. Discuss how the Caribbean's history is reflected in its culture today.

2. How does the legacy of the colonial past continue in Central America?

3. Why has there been a high rate of emigration from some of the Caribbean countries?

4. Discuss the relationship between Puerto Rico and the United States.

Name ______________________ Class ______________ Date ____________

Chapter Test Form A

South America

REVIEWING FACTS

MATCHING *(3 points each)* In the space provided, write the letter of the term or place that matches each description. Some answers will not be used.

_____ **1.** Large estates on the Pampas

_____ **2.** A chain of high plateaus

_____ **3.** Small farms that are often unprofitable

_____ **4.** The line of elevation above which trees do not grow

_____ **5.** A tropical fruit with starchy roots

_____ **6.** A Southern Common Market

_____ **7.** A large plains area between Colombia and Venezuela

_____ **8.** A small country between two larger, more powerful countries

_____ **9.** Rock or sand layers that contain oil

_____ **10.** Large urban slums

a. tree line
b. Paraná
c. latifundia
d. coup
e. favelas
f. tepuís
g. manioc
h. minifundia
i. Mercosur
j. Llanos
k. buffer state
l. tar sands

FILL IN THE BLANK *(3 points each)* For each of the following statements, fill in the blank with the appropriate word, phrase, or place.

1. After a war with ______________________, Bolivia was left landlocked.

2. The Amazon River is the world's largest ______________________ climate region.

3. The largest percentage of South American Indians lives in

______________________.

4. Chile is the world's largest producer and exporter of ______________________.

5. The Guaraní are a South American Indian people who lived in what is now eastern

______________________.

6. Vast deposits of oil are located around Lake ______________________.

7. Two of ______________________'s main products are coffee and cut flowers.

8. ______________________ conquered the Inca in the 1530s.

9. Uruguay is a buffer state between Argentina and ______________________.

10. The ______________________ were South American Indians living in the Colombian Andes who developed gold-working skills.

UNDERSTANDING IDEAS *(3 points each)* For each of the following, write the letter of the best choice in the space provided.

______ **1.** What is one factor that limits the profits of many industries in South America?
- **a.** strict employment regulations
- **b.** terrorism
- **c.** all parts and raw materials must be imported
- **d.** lack of educated workforce

______ **2.** Which of the following is not a major river in South America?
- **a.** Amazon
- **b.** Llanos
- **c.** Orinoco
- **d.** Paraná

______ **3.** What was South America's greatest early civilization?
- **a.** Guaraní
- **b.** Aztec
- **c.** Inca
- **d.** Chibcha

______ **4.** What country is famous for its emeralds?
- **a.** Venezuela
- **b.** Colombia
- **c.** Brazil
- **d.** Chile

______ **5.** What is a major cause of poverty and unrest in South America?
- **a.** lack of natural resources
- **b.** ethnic tensions
- **c.** lack of education
- **d.** inequality in land ownership

PRACTICING SKILLS *(5 points each)* Study the table and answer the questions that follow.

Brazil's Gross Domestic Product (GDP) per Capita by Region, 1997

Region	Population (in millions)	GDP (in Brazilian reals)
North	11.3	3,293
Northeast	44.8	2,494
Southeast	67.0	7,436
South	23.5	6,402
East Central	10.5	5,008
Brazil	157.1	5,413

1. Which region of Brazil is the poorest economically?

____

2. Which region of Brazil has the highest percentage of the total population?

____

COMPOSING AN ESSAY *(15 points)* On a separate sheet of paper, write an essay in response to *one* of the following.

1. How is the history of Brazilian independence different than that of the Spanish colonies in South America?

2. Discuss why the Amazon River is unique. Provide evidence to support this view.

Name ______________________ Class ______________ Date ____________

Chapter Test Form B

South America

SHORT ANSWER *(10 points each)* Provide brief answers for each of the following. Use examples to support your answers.

1. What created the Andes? How does tectonic activity affect the region today?

2. Why does the continent of South America have a wide variety of climate regions?

3. Why does Patagonia have semiarid and arid climates?

4. Describe how terrorism is affecting life in Colombia.

5. What are four developments of Inca civilization that are still visible today?

PRACTICING SKILLS *(10 points each)* Study the table and answer the questions that follow.

Brazil's Gross Domestic Product (GDP) per Capita by Region, 1997

Region	Population (in millions)	GDP (in Brazilian reals)
North	11.3	3,293
Northeast	44.8	2,494
Southeast	67.0	7,436
South	23.5	6,402
East Central	10.5	5,008
Brazil	157.1	5,413

1. What two regions have the closest GDP to the national average?

2. Why do you think that the Northern region has the smallest population?

COMPOSING AN ESSAY *(30 points)* On a separate sheet of paper, write an essay in response to *one* of the following.

1. Why did the South American Indian population fall sharply during the colonial period?

2. Describe the settlement patterns of South America.

3. How did the independence movements in South America affect people's lives?

4. What is the purpose of Mercosur? What countries are full members in Mercosur?

Name ______________________ Class ______________ Date ______________

Unit Test Form A

Middle and South America

REVIEWING FACTS

MATCHING *(3 points each)* In the space provided, write the letter of the term or place that matches each description. Some answers will not be used.

_____ **1.** A narrow strip of land that connects two larger land areas

_____ **2.** Tree whose roots grow in salt water

_____ **3.** Crops grown for sale in a market

_____ **4.** A small country situated between two larger, more powerful countries

_____ **5.** Large urban slums

_____ **6.** A term that means "native"

_____ **7.** Factories in Mexico that produce goods for U.S. markets

_____ **8.** A chain of high plateaus

_____ **9.** People of African and European descent

_____ **10.** People of mixed European and Indian heritage

a. isthmus
b. creole
c. tar sands
d. mestizos
e. buffer state
f. indigenous
g. favelas
h. mulattos
i. cash crops
j. tepuís
k. mangrove
l. maquiladoras

FILL IN THE BLANK *(3 points each)* For each of the following statements, fill in the blank with the appropriate word, phrase, or place.

1. ______________________ is the only non-Spanish speaking country in Central America.

2. The ______________________ separates the Gulf of Mexico and the Pacific Ocean.

3. Chile is the world's largest producer and exporter of ______________________.

4. The ______________________ were South American Indians living in the Colombian Andes who developed gold-working skills.

5. One result of the Mexican Revolution was ______________________ reform.

6. One of Central America's greatest natural resources is ______________________, the basis for its tourism industry.

7. Mexico's poor ______________________ is a major obstacle to continued economic growth.

8. The Panama Canal was controlled by the ______________________ until 1999.

9. The Amazon River is the world's largest ______________________ climate region.

10. ______________________ is a style of cultivation in which several crops are grown together in a single plot.

UNDERSTANDING IDEAS *(3 points each)* For each of the following, write the letter of the best choice in the space provided.

______ **1.** What is the most industrialized country in the Caribbean?
- **a.** Cuba
- **b.** Dominican Republic
- **c.** Puerto Rico
- **d.** Bahamas

______ **2.** What was the capital city of the Aztecs?
- **a.** Mazatlán
- **b.** Tenochtitlán
- **c.** Campeche
- **d.** Tampico

______ **3.** What is one limitation of many industries in South America?
- **a.** strict employment regulations
- **b.** terrorism
- **c.** all parts and raw materials must be imported
- **d.** lack of educated workforce

______ **4.** Which of the following is not a major river in South America?
- **a.** Amazon
- **b.** Llanos
- **c.** Orinoco
- **d.** Paraná

______ **5.** Which island was formed from volcanoes?
- **a.** Martinique
- **b.** Bahamas
- **c.** Hispaniola
- **d.** Cuba

PRACTICING SKILLS *(5 points each)* Study the table and answer the questions that follow.

Brazil's Gross Domestic Product (GDP) per Capita by Region, 1997

Region	Population (in millions)	GDP (in Brazilian reals)
North	11.3	3,293
Northeast	44.8	2,494
Southeast	67.0	7,436
South	23.5	6,402
East Central	10.5	5,008
Brazil	157.1	5,413

1. What two regions have the closest GDP to the national average?

____

2. Which region of Brazil is the poorest?

____

COMPOSING AN ESSAY *(15 points)* On a separate sheet of paper, write an essay in response to *one* of the following.

1. Discuss how daily life in Mexico has changed with economic growth.

2. What are the benefits of promoting ecotourism as an important part of Costa Rica's economy?

Name ______________________ Class ______________ Date ____________

Unit Test Form B

Middle and South America

SHORT ANSWER *(10 points each)* Provide brief answers for each of the following. Use examples to support your answers.

1. Why does the continent of South America have a wide variety of climate regions?
2. What physical process is responsible for creating many islands that are found in the Lesser Antilles?
3. Describe how terrorism is affecting life in Colombia.
4. What factors make the Yucatán Peninsula prone to sinkholes?
5. Explain why the term West Indies is used to describe the islands of the Caribbean region.

PRACTICING SKILLS *(10 points each)* Study the graph and answer the questions that follow.

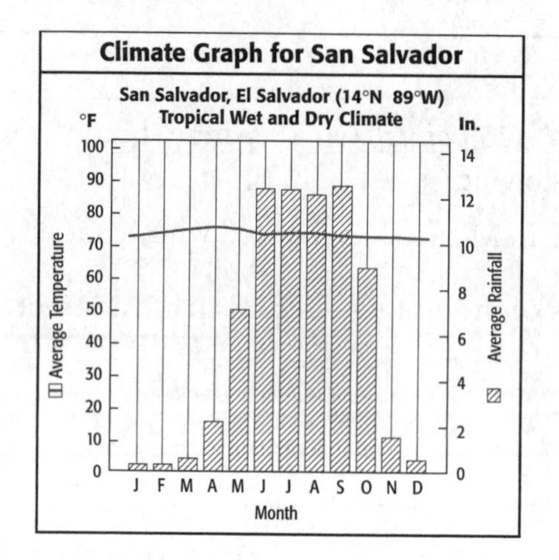

1. In what months does San Salvador receive the least rain?
2. What is the approximate annual rainfall of San Salvador?

COMPOSING AN ESSAY *(30 points)* On a separate sheet of paper, write an essay in response to *one* of the following.

1. Discuss the landforms that influence Mexico's climate. Why is there such a broad range of climates in the country?
2. Why has there been a high rate of emigration from some of the Caribbean countries?
3. Describe how a typical colonial town in Mexico evolved and was organized.
4. How did the independence movements in South America affect peoples' lives?

Name ______________________ Class ______________ Date ____________

Chapter Test Form A

Natural Environments of Europe

REVIEWING FACTS

MATCHING *(3 points each)* In the space provided, write the letter of the term or place that matches each description. Some answers will not be used.

_____ **1.** Major European river that empties into the North Sea

_____ **2.** Marks the eastern boundary of Europe

_____ **3.** Fertile, fine-grain, windblown soil

_____ **4.** An extremely fertile valley and major farming center

_____ **5.** Location of major fishing grounds

_____ **6.** Major European river that empties in to the Black Sea

_____ **7.** Earthen walls built to hold back water

_____ **8.** This region has a marine west coast climate

_____ **9.** Warm ocean current that flows along northwestern Europe

_____ **10.** Lands reclaimed from the sea

a. North Atlantic Drift
b. Po Valley
c. Ural Mountains
d. dikes
e. Danube River
f. British Isles
g. polders
h. loess
i. Central Uplands
j. North Sea
k. Rhine River
l. Scandinavian Peninsula

FILL IN THE BLANK *(3 points each)* For each of the following statements, fill in the blank with the appropriate word, phrase, or place.

1. Europe's ______________________ and good natural harbors have made trade easy.

2. The port of Murmansk stays ice free in the winter due to the ______________________.

3. Oil and natural gas were discovered beneath the ______________________ Sea in the 1960s.

4. Wind-deposited ______________________ soils can keep their fertility for years.

5. Both Paris and Berlin are located on the ______________________.

6. Germany's Ruhr Valley contains one of the world's largest

______________________ deposits.

7. Europe relies heavily on ______________________ imports.

8. The Guadalquivir River valley contains one of Europe's major

______________________ regions.

9. Avalanches are common in winter in the ______________________.

10. Most of Europe lies within a ______________________ biome.

UNDERSTANDING IDEAS *(3 points each)* For each of the following, write the letter of the best choice in the space provided.

______ **1.** What is the chief characteristic of a navigable river?
- **a.** It is deep enough and wide enough for shipping.
- **b.** It does not empty into a lake.
- **c.** It more than one mile wide.
- **d.** It is more than one half mile wide.

______ **2.** Which of the following is not a major fishing country?
- **a.** Ireland
- **b.** Iceland
- **c.** Spain
- **d.** Denmark

______ **3.** Which of the following is not a contributing factor to the extinction of some species in Europe?
- **a.** growth of urban areas
- **b.** prolonged drought
- **c.** pollution
- **d.** loss of habitat

______ **4.** How is the Mediterranean Sea connected to the Black Sea?
- **a.** through the Canal du Midi
- **b.** through the Bosporus
- **c.** through the Rhine River
- **d.** through the Baltic Sea

______ **5.** Which of the following is not part of the Alpine Mountain System?
- **a.** Jura Mountains
- **b.** Carpathian Mountains
- **c.** Apennines
- **d.** Pyrenees

PRACTICING SKILLS *(5 points each)* Study the table and answer the questions that follow.

Annual Fish Catches

Country	1986	1991	1997
Denmark	1.5	1.75	1.8
Iceland	1.6	1.0	2.25
Norway	1.75	2.25	2.8
Spain	1.25	1.25	1.15

NOTE: Data are in metric tons of fish (in millions). Data for Denmark is for 1987, 1992, and 1997.
Source: World Almanac and Book of Facts

1. Which country has stayed the most constant in the amount of fish caught during these time periods?

____

2. Which country had the smallest total catch for the years listed?

____

COMPOSING AN ESSAY *(15 points)* On a separate sheet of paper, write an essay in response to *one* of the following.

1. Why is it that much of Europe has mild temperatures compared to world regions of similar latitude?

2. What has been the result of the change from coal to oil and natural gas as a major energy resource in Europe?

Name ______________________ Class ______________ Date ______________

Chapter Test Form B

Natural Environments of Europe

SHORT ANSWER *(10 points each)* Provide brief answers for each of the following. Use examples to support your answers.

1. How did glaciers affect the landscapes of Europe?

2. What are Europe's three major climate types?

3. What are the major landforms within the Central Uplands?

4. Name two river valleys that are major farming centers.

5. What trees and animals are found in the temperate forest biome?

PRACTICING SKILLS *(10 points each)* Study the table and answer the questions that follow.

Annual Fish Catches

Country	1986	1991	1997
Denmark	1.5	1.75	1.8
Iceland	1.6	1.0	2.25
Norway	1.75	2.25	2.8
Spain	1.25	1.25	1.15

NOTE: Data are in metric tons of fish (in millions). Data for Denmark is for 1987, 1992, and 1997.
Source: World Almanac and Book of Facts

1. What two countries had the highest catches in 1997?

2. What country has caught the most fish in total?

COMPOSING AN ESSAY *(30 points)* On a separate sheet of paper, write an essay in response to *one* of the following.

1. How have human activities affected Europe's plants and wildlife?

2. Why does Europe enjoy mild climates, given its high latitude?

3. What role does fishing play in the European economy?

4. Why is Europe ideally located for trade by sea?

Name ______________________ Class ______________ Date ______________

Chapter Test Form A

Northern and Western Europe

REVIEWING FACTS

MATCHING *(3 points each)* In the space provided, write the letter of the term or place that matches each description. Some answers will not be used.

_____ **1.** Unable to support human life

_____ **2.** First-ranked and dominant city in a country

_____ **3.** Settlement by successive groups of people with distinctive cultures

_____ **4.** Economic system in which government owns the means of producing goods

_____ **5.** Northern coastal region of Belgium

_____ **6.** Economic and political cooperative organization

_____ **7.** Characterized by many foreign influences

_____ **8.** A city in Northern Ireland

_____ **9.** A primate city

_____ **10.** Industries owned and operated by the government

a. geysers
b. European Union
c. cosmopolitan
d. socialism
e. sequent occupance
f. Belfast
g. Paris
h. uninhabitable
i. Corsica
j. primate city
k. nationalized
l. Flanders

FILL IN THE BLANK *(3 points each)* For each of the following statements, fill in the blank with the appropriate word, phrase, or place.

1. The largest immigrant population in France is from ______________________.

2. In terms of religion, almost all Scandinavians are ______________________.

3. ______________________ is second only to the United States in agricultural exports.

4. Greenland is a self-governing territory of ______________________.

5. The ______________________ drove the Celtic peoples of lowland Britain to highland Britain.

6. A common ______________________ and ______________________ have created a strong cultural identity in France.

7. Iceland has tremendous geothermal energy as evidenced by the many ______________________ found there.

8. ______________________ has the highest per capita GDP in the world.

9. The ______________________ in Ireland caused large numbers of people to immigrate to the United States.

10. The two regions of Belgium are ______________________ and ______________________.

UNDERSTANDING IDEAS *(3 points each)* For each of the following, write the letter of the best choice in the space provided.

______ **1.** What group was the last to invade and occupy Great Britain?
- **a.** Normans
- **b.** Vikings
- **c.** Angles
- **d.** Celts

______ **2.** The large urban area of the Netherlands is known as
- **a.** the Randstad.
- **b.** the polder.
- **c.** the EU.
- **d.** the Hague.

______ **3.** Lapland stretches across which three Scandinavian countries?
- **a.** Norway, Denmark, and Sweden
- **b.** Denmark, Sweden, and Finland
- **c.** Iceland, Norway, and Finland
- **d.** Norway, Sweden, and Finland

______ **4.** What are the Benelux countries?
- **a.** Belgium, France, and the Netherlands
- **b.** Belgium, Luxembourg, and France
- **c.** the Netherlands, Switzerland, and Luxembourg
- **d.** Belgium, Luxembourg, and the Netherlands

______ **5.** What river runs through the middle of Paris?
- **a.** Thames
- **b.** Po
- **c.** Seine
- **d.** Loire

PRACTICING SKILLS *(5 points each)* Study the diagram and answer the questions that follow.

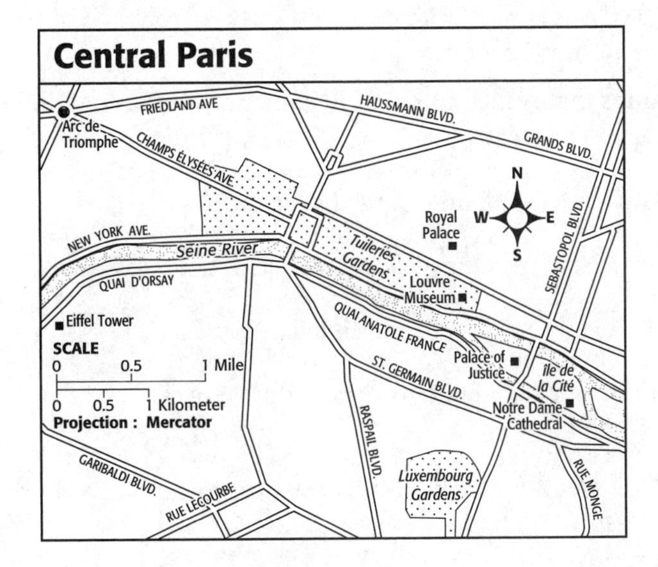

1. Where is Notre Dame located?

____

____

2. What point of interest is farthest from the Luxembourg Gardens?

____

____

COMPOSING AN ESSAY *(15 points)* On a separate sheet of paper, write an essay in response to *one* of the following.

1. What has been the impact of government control on the economies of Western and Northern Europe?

2. Discuss the causes of the continuing violence in Northern Ireland.

Name ______________________ Class ______________ Date ______________

Chapter Test Form B

Northern and Western Europe

SHORT ANSWER *(10 points each)* Provide brief answers for each of the following. Use examples to support your answers.

1. Name three groups that settled in Great Britain and contributed to its distinctive cultural landscape.

2. How has the economy of Ireland changed over time?

3. How have France's large overseas colonies affected its population?

4. Name the countries that make up Scandinavia.

5. Why are the countries of the Benelux grouped together?

PRACTICING SKILLS *(10 points each)* Study the diagram and answer the questions that follow.

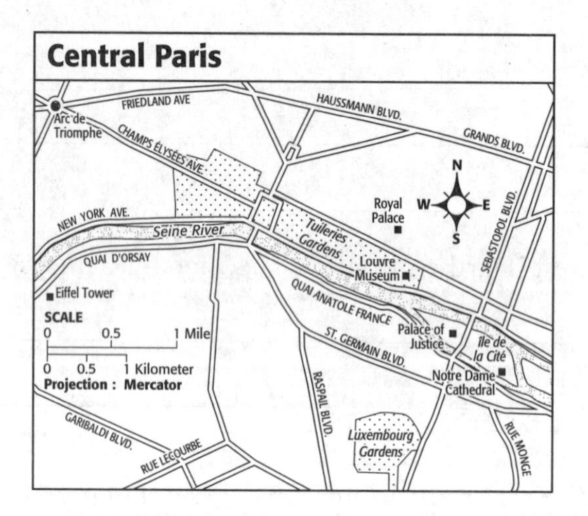

1. What street connects the Arc de Triomphe and the Tuileries Gardens?

2. What gardens are located south of the Seine?

COMPOSING AN ESSAY *(30 points)* On a separate sheet of paper, write an essay in response to *one* of the following.

1. Compare and contrast the cultures of Ireland and Great Britain.

2. Describe the economy of France.

3. Describe the settlement patterns of the Scandinavian countries.

4. How did the Industrial Revolution affect the position of Great Britain in world politics?

Name ______________________ Class ______________ Date ______________

Chapter Test Form A

Central Europe

REVIEWING FACTS

MATCHING *(3 points each)* In the space provided, write the letter of the term or place that matches each description. Some answers will not be used.

_____ **1.** The capital and transportation hub of Poland

_____ **2.** A group of states joined together for a common purpose

_____ **3.** An agreement between countries to support one another against enemies

_____ **4.** The combining of two areas with different activities and strengths

_____ **5.** Capital city of Austria

_____ **6.** Section of a city where a minority group is forced to live

_____ **7.** Largest city in Switzerland and world-banking center

_____ **8.** The largest German state in area

_____ **9.** An area separated from the rest of a country by the territory of another country

_____ **10.** A Swiss state

a. Prague
b. canton
c. Bavaria
d. Warsaw
e. complementary region
f. Zürich
g. Budapest
h. ghetto
i. Vienna
j. alliance
k. confederation
l. exclave

FILL IN THE BLANK *(3 points each)* For each of the following statements, fill in the blank with the appropriate word, phrase, or place.

1. About 90 percent of ______________________'s population is Magyar.

2. ______________________ is Poland's main seaport.

3. The first ruler to unite several German kingdoms was ______________________.

4. ______________________ is a neutral country.

5. The Balts and the ______________________ were the two groups of people who originally settled in Latvia, Lithuania, and Estonia.

6. ______________________ has the largest GDP in Central Europe and the fourth-largest in the world.

7. The capital city of Slovakia is ______________________.

8. The capital of Switzerland is ______________________.

9. The Habsburgs were the most powerful family in ______________________.

10. The breakup of Czechoslovakia into the Czech Republic and Slovakia is called the

______________________.

UNDERSTANDING IDEAS *(3 points each)* For each of the following, write the letter of the best choice in the space provided.

______ **1.** What is a cymbaly?
a. a type of percussion instrument popular in the Baltic countries
b. a type of Viennese dessert
c. the currency of Estonia
d. a dialect of Polish spoken in Latvia

______ **2.** The first trading group in Germany was
a. an alliance with Switzerland.
b. in Prussia.
c. the Hanseatic League.
d. in Bavaria.

______ **3.** Which city in Switzerland is a transportation center on the Rhine?
a. Zürich
b. Bern
c. Basel
d. Geneva

______ **4.** Which two countries form a complementary region?
a. the Czech Republic and Slovakia
b. Latvia and Lithuania
c. Germany and Austria
d. Estonia and Kaliningrad

______ **5.** Which country has a rich history of folk and music traditions?
a. Slovakia
b. Hungary
c. Lithuania
d. All of the above

Name ______________________ Class ______________ Date ____________

PRACTICING SKILLS *(5 points each)* Study the maps and answer the questions that follow.

1. How did the size of the kingdom of Hungary compare to present-day Hungary?

____

2. Which country seems to have gained the most land since the collapse of the Austro-Hungarian Empire?

____

COMPOSING AN ESSAY *(15 points)* On a separate sheet of paper, write an essay in response to *one* of the following.

1. How has the Polish economy changed since the end of the Communist era?

2. What economic and political problems helped bring Germany's Nazi Party to power in 1933?

Name ______________________ Class ______________ Date ____________

Chapter Test Form B

Central Europe

SHORT ANSWER *(10 points each)* Provide brief answers for each of the following. Use examples to support your answers.

1. What makes Vienna, Austria, an important city in Central Europe?

2. What countries make up the Baltic region? Why is trade essential to these countries?

3. Why did the population of Budapest decline in the 1990s?

4. Describe Switzerland's cultural diversity.

5. Explain why loess is important to Poland's economy.

PRACTICING SKILLS *(10 points each)* Study the maps and answer the questions that follow.

1. What other countries have occupied the land that is now Poland?

2. What countries were created as a result of World War I?

COMPOSING AN ESSAY *(30 points)* On a separate sheet of paper, write an essay in response to *one* of the following.

1. Describe the culture of Germany.

2. How did Kaliningrad become an exclave of Russia?

3. Compare and contrast the economies of the Czech Republic and Slovakia.

4. How is Hungary different from other countries in Central Europe?

Name ______________________ Class ______________ Date ______________

Chapter Test Form A

Southern Europe and the Balkans

REVIEWING FACTS

MATCHING *(3 points each)* In the space provided, write the letter that matches each description. Choose your answers from the list below. Some answers will not be used.

_____ **1.** A large island off the coast of Italy

_____ **2.** Extremely small, autonomous countries

_____ **3.** The main port of Spain

_____ **4.** An industrial city in northern Italy

_____ **5.** Bark stripped from the trunks of certain trees

_____ **6.** The largest city in the Balkans

_____ **7.** An area that is completely surrounded by another region

_____ **8.** Portugal's largest city and capital

_____ **9.** A self-governing city and the surrounding area

_____ **10.** A period of renewed interest in the arts and architecture

a. Barcelona
b. Sardinia
c. cork
d. Bucharest
e. Belgrade
f. microstates
g. enclave
h. Lisbon
i. autonomy
j. Renaissance
k. city-state
l. Turin

FILL IN THE BLANK *(3 points each)* For each of the following statements, fill in the blank with the appropriate word, phrase, or place.

1. Sicily and Sardinia are both parts of ______________________.

2. The dictator ______________________ ruled Spain for over 35 years.

3. Early systems of democratic government originated in ______________________.

4. The Italian dictator ______________________ formed an alliance with Germany during World War II.

5. Bosnia and Herzegovina, ______________________, Macedonia, and Slovenia are all countries that were once part of Yugoslavia.

6. Portugal still retains control of its colonies in ______________________ and ______________________.

7. The Iberian Peninsula was occupied by the ______________________ for over 700 years.

8. The microstate of ____________________ lies between France and Spain.

9. The ETA is a group of ____________________ separatists who are fighting for complete autonomy from Spain.

10. ____________________ is the main seaport of Portugal.

UNDERSTANDING IDEAS *(3 points each)* For each of the following, write the letter of the best choice in the space provided.

______ **1.** The Moors brought all the following except _____ to Spain.
- **a.** the Islamic faith
- **b.** new road building techniques
- **c.** new crops
- **d.** new irrigation techniques

______ **2.** The "breadbasket" of Italy is centered on
- **a.** southern Italy.
- **b.** the Mezzogiorno.
- **c.** the Po River.
- **d.** northern Italy.

______ **3.** Africa lies just 8 kilometers away from Spain across what body of water?
- **a.** Adriatic Sea
- **b.** Atlantic Ocean
- **c.** Strait of Gibraltar
- **d.** Bay of Biscay

______ **4.** Spain is famous for producing all of these agricultural products except
- **a.** oranges
- **b.** olive oil
- **c.** wine
- **d.** wheat

______ **5.** The eastern Balkans region is composed of what three countries?
- **a.** Bulgaria, Slovenia, and Kosovo
- **b.** Bulgaria, Romania, and Moldova
- **c.** Romania, Kosovo, and Croatia
- **d.** Moldova, Romania, and Serbia

PRACTICING SKILLS *(5 points each)* Study the diagram and answer the questions that follow.

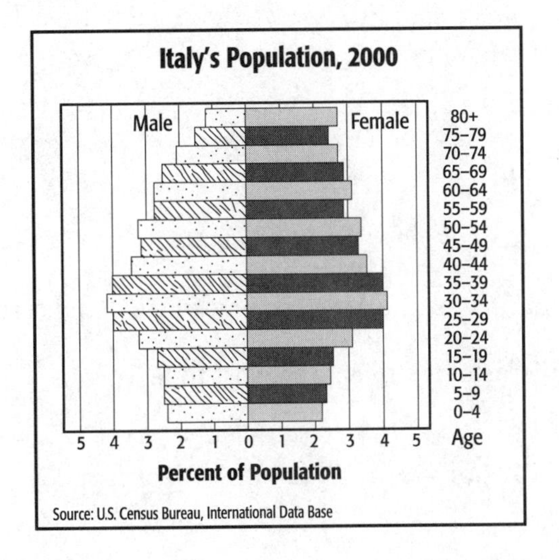

1. What age group currently represents the largest percentage of the population?

____

____

2. What age group has the greatest difference between male and female population?

____

____

COMPOSING AN ESSAY *(15 points)* On a separate sheet of paper, write an essay in response to *one* of the following.

1. Compare and contrast the eastern and western Balkans.

2. Briefly describe the history of Spain from A.D. 700 to 1975.

Name ____________________ Class ______________ Date ______________

Chapter Test Form B

Southern Europe and the Balkans

SHORT ANSWER *(10 points each)* Provide brief answers for each of the following. Use examples to support your answers.

1. How is Portugal similar to Spain?

2. What countries make up the western Balkans?

3. What are the major challenges for Greece today?

4. Why was trade important for the development of the Renaissance?

5. Why are Europe's microstates still able to exist today?

PRACTICING SKILLS *(10 points each)* Study the diagram and answer the questions that follow.

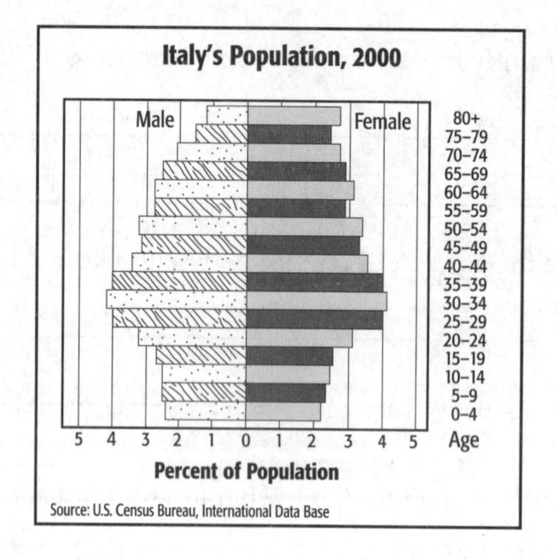

1. What does this chart indicate about Italy's population growth?

2. What age group represents the smallest percentage of the entire population?

COMPOSING AN ESSAY *(30 points)* On a separate sheet of paper, write an essay in response to *one* of the following.

1. What factors have contributed to continued fighting in the western Balkans?

2. Why are the Balkans so ethnically diverse compared with the rest of southern Europe?

3. How has the history of Spain and its empire had far-reaching effects?

4. How has Italy influenced the rest of Europe and the world?

Name ______________________ Class ______________ Date ____________

Unit Test Form A

Europe

REVIEWING FACTS

MATCHING *(3 points each)* In the space provided, write the letter of the term or place that matches each description. Some answers will not be used.

_____ **1.** The combining of two areas with different activities and strengths

_____ **2.** Marks the eastern boundary of Europe

_____ **3.** The largest city in the Balkans

_____ **4.** Economic system in which government owns the means of producing goods

_____ **5.** The capital and transportation hub of Poland

_____ **6.** Major European river that empties into the Black Sea

_____ **7.** Industries owned and operated by the government

_____ **8.** The main port of Spain

_____ **9.** Settlement by successive groups of people with distinctive cultures

_____ **10.** Major European River that empties into the North Sea

a. Ural Mountains
b. Barcelona
c. Warsaw
d. Danube River
e. sequent occupance
f. complementary region
g. Belgrade
h. socialism
i. Bucharest
j. nationalized
k. Rhine River
l. Prague

FILL IN THE BLANK *(3 points each)* For each of the following statements, fill in the blank with the appropriate word, phrase, or place.

1. The first ruler to unite several German kingdoms was ____________________.

2. ____________________ is second only to the United States in agricultural exports.

3. Oil and natural gas were discovered beneath the ____________________ Sea in the 1960s.

4. The Italian dictator ____________________ formed an alliance with Germany during World War II.

5. The two regions of Belgium are ____________________ and ____________________.

6. ______________________ has the largest GDP in Central Europe, the fourth-largest in the world.

7. Most of Europe lies within a ______________________ biome.

8. Bosnia and Herzegovina, ______________________, Macedonia, and Slovenia are all countries that were once part of Yugoslavia.

9. The Guadalquivir River valley contains one of Europe's major ______________________ regions.

10. The microstate of ______________________ lies between France and Spain.

UNDERSTANDING IDEAS *(3 points each)* For each of the following, write the letter of the best choice in the space provided.

_____ **1.** Which of the following is not part of the Alpine Mountain System?
a. Jura Mountains
b. Carpathian Mountains
c. Apennines
d. Pyrenees

_____ **2.** The "breadbasket" of Italy is centered on what region?
a. southern Italy
b. the Mezzogiorno
c. the Po River
d. northern Italy

_____ **3.** What group was the last to invade and occupy Great Britain?
a. Normans
b. Vikings
c. Angles
d. Celts

_____ **4.** The first trading group in Germany was
a. an alliance with Switzerland.
b. in Prussia.
c. the Hanseatic League.
d. in Bavaria.

_____ **5.** Which of the following is not a contributing factor to the extinction of some species in Europe?
a. growth of urban areas
b. prolonged drought
c. pollution
d. loss of habitat

PRACTICING SKILLS *(5 points each)* Study the map and answer the questions that follow.

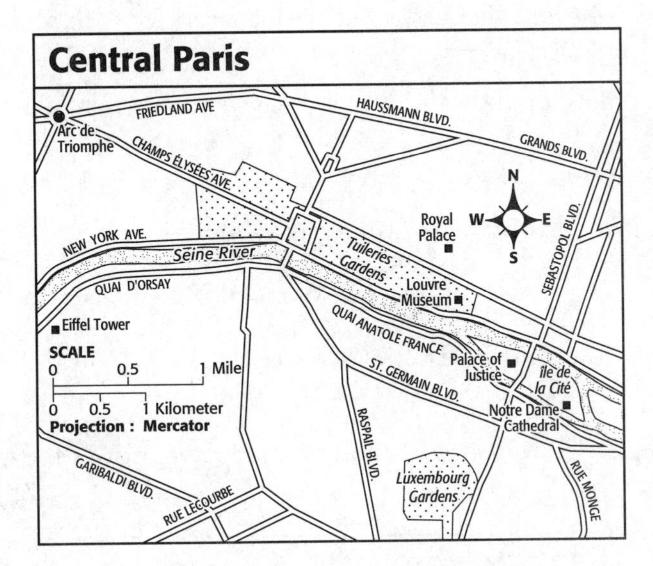

1. Where is Notre Dame located?

____

2. What point of interest is farthest from the Luxembourg Gardens?

____

COMPOSING AN ESSAY *(15 points)* On a separate sheet of paper, write an essay in response to *one* of the following.

1. Why does much of Europe have mild temperatures compared to other regions of the world at similar latitudes?

2. How has the Polish economy changed since the end of the Communist era?

Name ______________________ Class ______________ Date ____________

Unit Test Form B

Europe

SHORT ANSWER *(10 points each)* Provide brief answers for each of the following. Use examples to support your answers.

1. What countries make up the Baltic region? Why is trade essential to these countries?

2. What are the major challenges for Greece today?

3. Name the countries that make up Scandinavia.

4. Why was trade so important for the development of the Renaissance?

5. How did glaciers affect the landscapes of Europe?

PRACTICING SKILLS *(10 points each)* Study the map and answer the questions that follow.

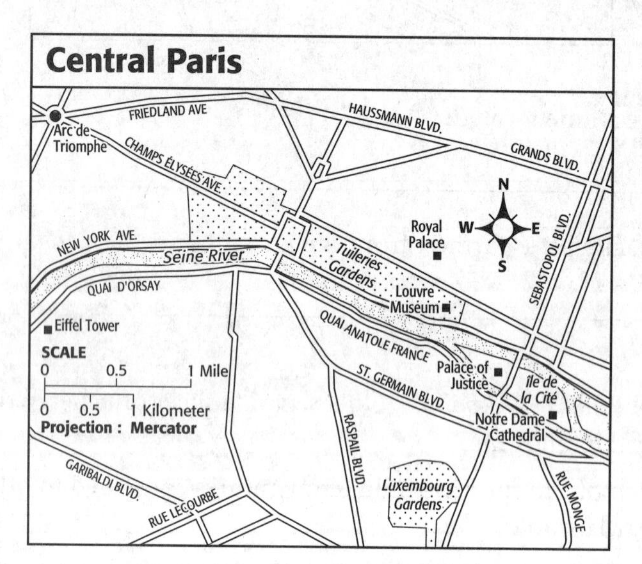

1. What street connects the Arc de Triomphe and the Tuileries Gardens?

2. What gardens are located south of the Seine?

COMPOSING AN ESSAY *(30 points)* On a separate sheet of paper, write an essay in response to *one* of the following.

1. Why is the Balkans so ethnically diverse compared with the rest of Southern Europe?

2. How did the Industrial Revolution affect the position of Great Britain in world politics?

3. Why is Europe ideally located for trade by sea?

4. Compare and contrast the economies of the Czech Republic and Slovakia.

Chapter Test Form A

Russia, Ukraine, and Belarus

REVIEWING FACTS

MATCHING *(3 points each)* In the space provided, write the letter of the term or place that matches each description. Some answers will not be used.

_____ **1.** Network of labor camps

_____ **2.** The production of consumer goods

_____ **3.** Area still under contention between Russia and Japan

_____ **4.** Ships that can break up ice in frozen waterways

_____ **5.** Factories that produce metal ores

_____ **6.** To resign

_____ **7.** City located on the Sea of Japan

_____ **8.** Forest region composed of evergreen trees

_____ **9.** Zones of frequent boundary changes and conflicts

_____ **10.** Location of Russia's first geothermal power station

a. autarky
b. smelters
c. abdicate
d. Lake Baikal
e. taiga
f. gulag
g. light industry
h. icebreakers
i. Vladivostok
j. Kuril Islands
k. shatter belts
l. Kamchatka

FILL IN THE BLANK *(3 points each)* For each of the following statements, fill in the blank with the appropriate word, phrase, or place.

1. Russia's most important economic region is located around

_______________________.

2. The eastward expansion of Russia was accomplished with the help of the

_______________________.

3. The part of Russia located east of the Ural Mountains is known as

_______________________.

4. _______________________ lies off the eastern coast of Siberia in the Sea of Okhotsk.

5. The main religion of the region is _______________________.

6. The first leader of the Soviet Union was _______________________.

7. Ukraine's heavy industry is located in the ______________________.

8. Soviet economic planners established a policy of ______________________, which stated that the country would itself produce all the goods it needed.

9. Russia's ______________________ region is one of the world's major grain-producing areas.

10. The first czar of all Russia was ______________________.

UNDERSTANDING IDEAS *(3 points each)* For each of the following, write the letter of the best choice in the space provided.

______ **1.** Which of the following has not occurred with the move to a market economy?
- **a.** increased civil unrest and rioting
- **b.** increased unemployment
- **c.** increased crime
- **d.** decrease in public health care

______ **2.** Where can large oil reserves be found?
- **a.** Arctic coastline
- **b.** Volga River basin
- **c.** Caspian Sea area
- **d.** Lake Baikal

______ **3.** What is borscht?
- **a.** a type of tea
- **b.** a type of grain
- **c.** a type of plant
- **d.** a type of soup

______ **4.** What contributes to Siberia's harsh winters?
- **a.** low elevations
- **b.** moderating ocean winds cannot reach inland
- **c.** high elevation blocks precipitation
- **d.** permafrost never melts

______ **5.** Who were the Rus?
- **a.** Scandinavian traders
- **b.** Mongols from the south
- **c.** Cossack warriors
- **d.** Siberian serfs

Name ______________________ Class ______________ Date ______________

PRACTICING SKILLS *(5 points each)* Study the map and answer the questions that follow.

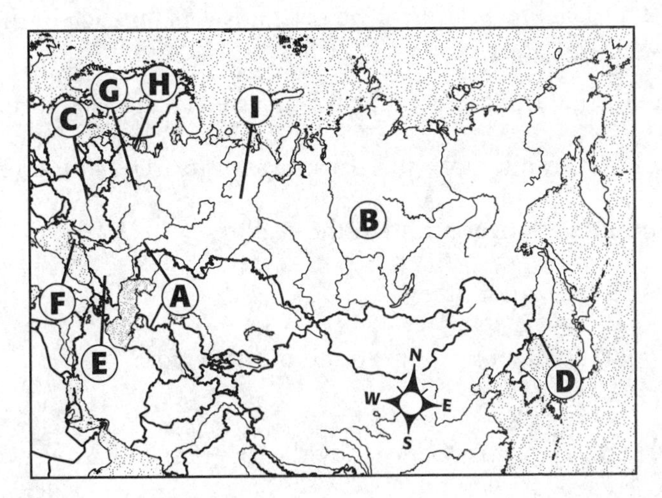

1. What letter represents Vladivostok?

____

2. What letter represents the Crimean Peninsula?

____

COMPOSING AN ESSAY *(15 points)* On a separate sheet of paper, write an essay in response to *one* of the following.

1. Why might a communist country want to follow a policy of autarky?

2. Discuss why greater Moscow is considered Russia's most important economic region.

Name ______________________ Class ______________ Date ____________

Chapter Test Form B

Russia, Ukraine, and Belarus

SHORT ANSWER *(10 points each)* Provide brief answers for each of the following. Use examples to support your answers.

1. Identify some Russian traditions that have survived changes in government.

2. Why is it that Russia cannot take full advantage of its rich natural resources?

3. Name three mountain ranges found in the region.

4. What are Russia's major resources?

5. What are the major religious groups found in the region?

PRACTICING SKILLS *(10 points each)* Study the diagram and answer the questions that follow.

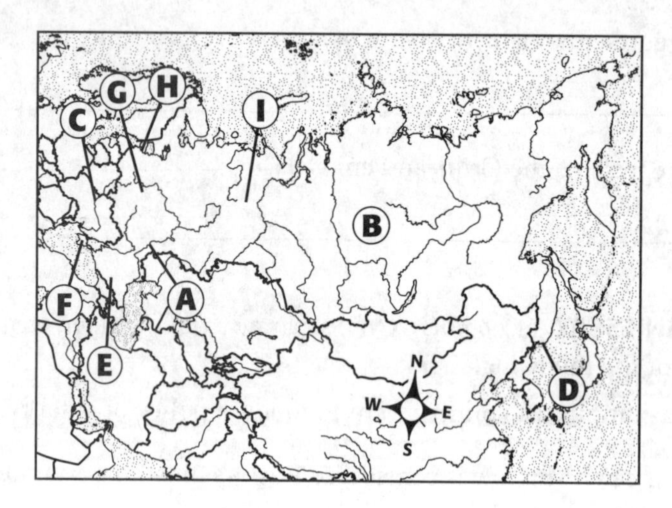

1. What letter represents St. Petersburg?

2. What letter represents the Central Siberian Plain?

COMPOSING AN ESSAY *(30 points)* On a separate sheet of paper, write an essay in response to *one* of the following.

1. Do you think the Soviet Union's command economy was successful? Why or why not?

2. How has society changed in Russia since the fall of the Soviet Union?

3. How did the Volga River basin become Russia's industrial heartland?

4. What basic change is occurring in the region's industry?

Name ______________________ Class ______________ Date ______________

Chapter Test Form A

Central Asia

REVIEWING FACTS

MATCHING *(3 points each)* In the space provided, write the letter of the term or place that matches each description. Some answers will not be used.

_____ **1.** Practice of moving livestock from summer pastures in the mountains to lowland pastures in winter

_____ **2.** A mountain range in northeast Central Asia

_____ **3.** Reliance on rainfall instead of irrigation in farming

_____ **4.** Cultivation of a single crop

_____ **5.** Central Asia's largest city

_____ **6.** The new capital of Kazakhstan

_____ **7.** River that begins in the Tian Shan and empties into the Aral Sea

_____ **8.** Movable round houses made of wool felt mats

_____ **9.** Extremely fertile area in Central Asia

_____ **10.** Element important in metal processing

a. dryland farming
b. zinc
c. yurts
d. Syr Dar'ya
e. Bishkek
f. transhumance
g. Fergana Valley
h. Pamirs
i. Astana
j. Altay Shan
k. Tashkent
l. monoculture

FILL IN THE BLANK *(3 points each)* For each of the following statements, fill in the blank with the appropriate word, phrase, or place.

1. The loss of water to irrigation along the Amu Dar'ya and Syr Dar'ya is leading to the disappearance of the ______________________.

2. The most precious natural resource in Central Asia is ______________________.

3. The main religion of Central Asia is ______________________.

4. In Central Asia, only Turkmenistan has a region with a ______________________ climate.

5. Kazakhstan and ______________________ have had the most success switching to a market economy.

6. ______________________ conquered Central Asia and built a capital in Samarqand.

7. Life expectancy in Central Asia is ______________________ than the world average.

8. The ______________________ mountains separate Turkmenistan and Iran.

9. After Russia colonized Central Asia, ______________________ became a monoculture.

10. The dominant heritage in Central Asia is ______________________.

UNDERSTANDING IDEAS *(3 points each)* For each of the following, write the letter of the best choice in the space provided.

_____ **1.** What countries share the Fergana Valley?
a. Tajikistan and Uzbekistan
b. Kyrgyzstan, Uzbekistan, and Tajikistan
c. Kyrgyzstan, Turkmenistan, and Uzbekistan
d. Kazakhstan and Turkmenistan

_____ **2.** Where are the major water sources of Central Asia located?
a. Turkmenistan and Uzbekistan
b. Tajikistan and Kyrgyzstan
c. Kazakhstan and Uzbekistan
d. Tajikistan and Turkmenistan

_____ **3.** Which desert is known for its black sand?
a. Kara Kum
b. Kyzyl Kum
c. Almaty
d. Issy-Kul

_____ **4.** What is the Silk Road?
a. a major textile region
b. a short cut through the Pamirs
c. a trade route to China
d. a flat plain in the Fergana Valley

_____ **5.** Which of the following is not a factor in the slow development of the area's natural resources?
a. threat of earthquakes
b. outdated equipment
c. corruption
d. poor transportation links

Name ______________________ Class ______________ Date ______________

PRACTICING SKILLS *(5 points each)* Study the table and answer the questions that follow.

Ethnic Group							
Country	**Kazakh**	**Kyrgyz**	**Russian**	**Tajik**	**Turkmen**	**Uzbek**	**Other**
Kazakhstan	46%		34.7%				19.3%
Kyrgyzstan		52.4%	18%			12.9%	16.7%
Tajikistan			3.5%	64.9%		25%	6.6%
Turkmenistan			6.7%		77%	9.2%	7.1%
Uzbekistan			5.5%			80%	14.5%

1. In which country do Uzbeks constitute the largest ethnic minority?

____

2. Why might Kazakhstan have a larger Russian population than other Central Asian countries?

____

COMPOSING AN ESSAY *(15 points)* On a separate sheet of paper, write an essay in response to *one* of the following.

1. Describe in detail the major issues facing Central Asian countries today.

2. Discuss how the government of the Soviet Union tried to eliminate religion in Central Asia. In you discussion include what happened in Central Asia after the collapse of the Soviet Union.

Name ______________________ Class ______________ Date ____________

Chapter Test Form B

Central Asia

SHORT ANSWER *(10 points each)* Provide brief answers for each of the following. Use examples to support your answers.

1. Why are there few large cities in Central Asia?
2. List the major mountain ranges in Central Asia.
3. What changes did Timur bring to the region during his rule?
4. Why types of animals are commonly found in the mountains of Central Asia?
5. What have been the positive benefits of Soviet rule in Central Asia?

PRACTICING SKILLS *(10 points each)* Study the table and answer the questions that follow.

	Ethnic Group						
Country	**Kazakh**	**Kyrgyz**	**Russian**	**Tajik**	**Turkmen**	**Uzbek**	**Other**
Kazakhstan	46%		34.7%				19.3%
Kyrgyzstan		52.4%	18%			12.9%	16.7%
Tajikistan			3.5%	64.9%		25%	6.6%
Turkmenistan			6.7%		77%	9.2%	7.1%
Uzbekistan			5.5%			80%	14.5%

1. Which country has the least variation in its ethnic makeup?
2. Which country has the smallest Russian population?

COMPOSING AN ESSAY *(30 points)* On a separate sheet of paper, write an essay in response to *one* of the following.

1. Discuss the changes brought to Central Asia by the Russian railroad.
2. Explain how the Soviet emphasis on education and health care still affects Central Asia.
3. What have been some of the negative impacts of the international boundaries imposed by the Soviet Union in Central Asia?
4. What factors limit the ability of Central Asian countries to improve their economies?

Name ______________________ Class ______________ Date ____________

Unit Test Form A

Russia and Northern Eurasia

REVIEWING FACTS

MATCHING *(3 points each)* In the space provided, write the letter of the term or place that matches each description. Some answers will not be used.

______ **1.** Element important in metal processing

______ **2.** City located on the Sea of Japan

______ **3.** To resign

______ **4.** Extremely fertile area in Central Asia

______ **5.** Zones of frequent boundary changes and conflicts

______ **6.** Movable round houses made of wool felt mats

______ **7.** Network of labor camps

______ **8.** Central Asia's largest city

______ **9.** Factories that process metal ores

______ **10.** Cultivation of a single crop

a. Kamchatka Peninsula
b. smelters
c. zinc
d. monoculture
e. abdicate
f. Fergana Valley
g. Vladivostok
h. Bishkek
i. shatter belts
j. yurts
k. gulag
l. Tashkent

FILL IN THE BLANK *(3 points each)* For each of the following statements, fill in the blank with the appropriate word, phrase, or place.

1. The first czar of all Russia was ______________________.

2. The dominant heritage in Central Asia is ______________________.

3. Soviet economic planners established a policy of ______________________, which stated that the country would itself produce all the goods it needed.

4. ______________________ conquered Central Asia and built a capital in Samarqand.

5. Russia's ______________________ region is one of the world's major grain-producing areas.

6. The part of Russia east of the Ural Mountains is known as ______________________.

7. The most precious natural resource in Central Asia is ______________________.

8. Ukraine's heavy industry is located in the ______________________.

9. The first leader of the Soviet Union was ______________________.

10. The main religion of Central Asia is ______________________.

UNDERSTANDING IDEAS *(3 points each)* For each of the following, write the letter of the best choice in the space provided.

______ **1.** Which of the following has not occurred in Russia since it changed to a market economy?
- **a.** increased civil unrest and rioting
- **b.** increased unemployment
- **c.** increased crime
- **d.** decrease in public health care

______ **2.** What countries share the Fergana valley?
- **a.** Tajikistan and Uzbekistan
- **b.** Kyrgyzstan, Uzbekistan, and Tajikistan
- **c.** Kyrgyzstan, Turkmenistan, and Uzbekistan
- **d.** Kazakhstan and Turkmenistan

______ **3.** What is the Silk Road?
- **a.** a major textile region
- **b.** a short cut through the Pamirs
- **c.** a trade route to China
- **d.** a flat plain in the Fergana Valley

______ **4.** What contributes to Siberia's harsh winters?
- **a.** low elevations
- **b.** inability of moderating ocean winds to reach inland
- **c.** high elevation blocking precipitation
- **d.** permafrost never melts

______ **5.** Who were the Rus?
- **a.** Scandinavian traders
- **b.** Mongols from the South
- **c.** Cossack warriors
- **d.** Siberian serfs

PRACTICING SKILLS *(5 points each)* Study the table and answer the questions that follow.

Ethnic Group							
Country	**Kazakh**	**Kyrgyz**	**Russian**	**Tajik**	**Turkmen**	**Uzbek**	**Other**
Kazakhstan	46%		34.7%				19.3%
Kyrgyzstan		52.4%	18%			12.9%	16.7%
Tajikistan			3.5%	64.9%		25%	6.6%
Turkmenistan			6.7%		77%	9.2%	7.1%
Uzbekistan			5.5%			80%	14.5%

1. In which country do Uzbeks constitute the largest ethnic minority?

____

2. Why might Kazakhstan have a larger Russian population than other Central Asian countries?

____

COMPOSING AN ESSAY *(15 points)* On a separate sheet of paper, write an essay in response to *one* of the following.

1. Describe in detail the major issues facing Central Asian countries today.

2. Why might a communist country want to follow a policy of autarky?

Name ________________________ Class ______________ Date ____________

Unit Test Form B

Russia and Northern Eurasia

SHORT ANSWER *(10 points each)* Provide brief answers for each of the following. Use examples to support your answers.

1. List the major mountain ranges in Central Asia.

2. Why is it that Russia cannot take full advantage of its natural resources?

3. How did the Volga River basin become Russia's industrial heartland?

4. Why are there few big cities in Central Asia?

5. Identify some Russian traditions that have survived changes in government.

PRACTICING SKILLS *(10 points each)* Study the table and answer the questions that follow.

	Ethnic Group						
Country	**Kazakh**	**Kyrgyz**	**Russian**	**Tajik**	**Turkmen**	**Uzbek**	**Other**
Kazakhstan	46%		34.7%				19.3%
Kyrgyzstan		52.4%	18%			12.9%	16.7%
Tajikistan			3.5%	64.9%		25%	6.6%
Turkmenistan			6.7%		77%	9.2%	7.1%
Uzbekistan			5.5%			80%	14.5%

1. Which country has the least variation in its ethnic makeup?

2. Which country has the smallest Russian population?

COMPOSING AN ESSAY *(30 points)* On a separate sheet of paper, write an essay in response to *one* of the following.

1. Explain how the Soviet emphasis on education and health care still affects Central Asia today.

2. How has society changed in Russia since the fall of the Soviet Union?

3. What factors limit the ability of Central Asian countries to improve their economies?

4. Do you think the Soviet Union's command economy was successful? Why or why not?

Name ______________________ Class ______________ Date ______________

Chapter Test Form A

The Persian Gulf and Interior

REVIEWING FACTS

MATCHING *(3 points each)* In the space provided, write the letter of the term or place that matches each description. Some answers will not be used.

_____ **1.** Is becoming wider as the African and Arabian plates move apart

_____ **2.** Capital city of Iran

_____ **3.** A group of nomadic herders who live in Southwest Asia

_____ **4.** River that begins in a humid region and flows to a dry region

_____ **5.** A line of hereditary rulers

_____ **6.** Borders the Arabian Peninsula to the south

_____ **7.** Capital city of Saudi Arabia

_____ **8.** Government based on religious laws

_____ **9.** Shia Muslim religious leader of the highest order

_____ **10.** Branch of Islam in which imams are descendants of Muhammad

a. theocracy
b. Shi'ism
c. Arabian Sea
d. Red Sea
e. Riyadh
f. ayatollah
g. Tehran
h. exotic river
i. oasis
j. dynasty
k. Bedouins
l. Kabul

FILL IN THE BLANK *(3 points each)* For each of the following statements, fill in the blank with the appropriate word, phrase, or place.

1. ______________________ is the capital of Afghanistan.

2. The Persian Gulf region lies at the intersection of ______________________ tectonic plates.

3. The most common language in Iran is ______________________.

4. ______________________ is Islam's holiest city.

5. The ______________________ is a desert of red sand.

6. Iran's current form of government is a ______________________.

7. About 90 percent of all Muslims are ______________________ Muslims.

8. The ______________________ are found along Iran's western coast.

9. The ______________________ developed what became known as the golden age of Persian culture.

10. The region's only wet climates are produced by the ______________________.

UNDERSTANDING IDEAS *(3 points each)* For each of the following, write the letter of the best choice in the space provided.

______ **1.** What is the Taliban?
- **a.** an extremist group that once ruled Afghanistan
- **b.** the extremist ruling group in Iraq
- **c.** the book written by Muhammad
- **d.** the strict constitution of Iran

______ **2.** Which country is not located on the Arabian Peninsula?
- **a.** Qatar
- **b.** Kuwait
- **c.** Iraq
- **d.** United Arab Emirates

______ **3.** Which is not a basic theme in this region?
- **a.** oil
- **b.** democratic rule
- **c.** the role of Islam
- **d.** preservation of traditional leaders

______ **4.** Which mountains are located along the northeast border of Iran?
- **a.** Elburz
- **b.** Kopet-Dag
- **c.** Hindu Kush
- **d.** Zagros

______ **5.** What lies to the south of the Arabian Peninsula?
- **a.** Mediterranean Sea and Shatt al Arab
- **b.** Arabian Sea and Red Sea
- **c.** Persian Gulf and Shatt al Arab
- **d.** Arabian Sea and the Gulf of Aden

PRACTICING SKILLS *(5 points each)* Study the map and answer the questions that follow.

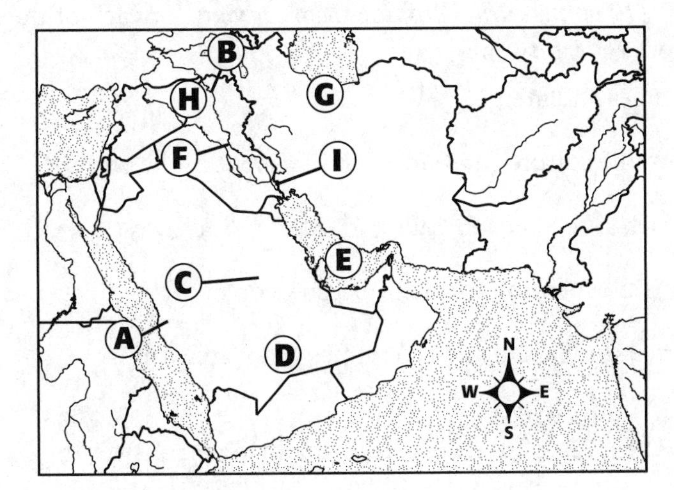

1. What city is identified by the letter A on the map?

____

2. What body of water is identified by the letter E on the map?

____

COMPOSING AN ESSAY *(15 points)* On a separate sheet of paper, write an essay in response to *one* of the following.

1. Discuss the importance of religion in the region.

2. Write a brief history of the region.

Name ________________ Class ________ Date ________

Chapter Test Form B

The Persian Gulf and Interior

SHORT ANSWER *(10 points each)* Provide brief answers for each of the following. Use examples to support your answers.

1. Why is the Red Sea widening?
2. How did the Arabic culture come to be so prevalent in the region?
3. List the countries of the Persian Gulf and interior Southwest Asia.
4. What is the current situation for the Kurdish people?
5. What are the major pillars of the economies of this region?

PRACTICING SKILLS *(10 points each)* Study the map and answer the questions that follow.

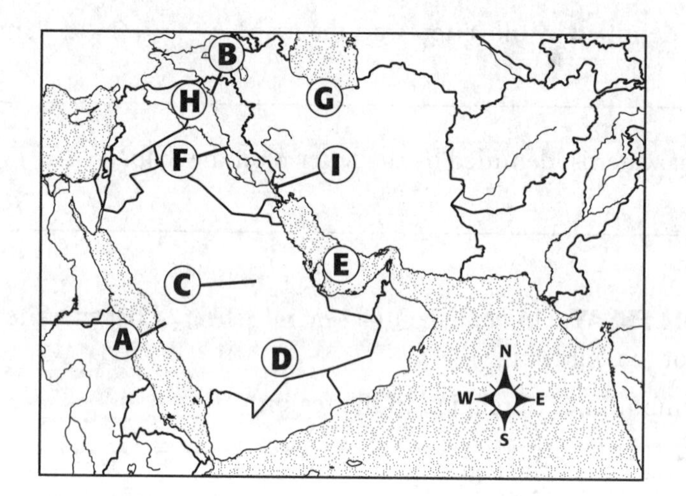

1. What landform is identified by the letter G?
2. What desert is located by the letter D?

COMPOSING AN ESSAY *(30 points)* On a separate sheet of paper, write an essay in response to *one* of the following.

1. Write a brief history of the region.
2. How does the availability of water in the region shape economic activity?
3. Why can this region be described as unstable politically, economically, and ecologically?
4. Why is the political stability of the region important to countries around the world?

Name ______________________ Class ______________ Date ____________

Chapter Test Form A

The Eastern Mediterranean

REVIEWING FACTS

MATCHING *(3 points each)* In the space provided, write the letter that matches each description. Choose your answers from the list below. Some answers will not be used.

_____ **1.** City in Lebanon once known as "the Paris of the Middle East"

_____ **2.** The largest ethnic minority in Turkey

_____ **3.** Territories placed under another country's control

_____ **4.** Large port city in Israel

_____ **5.** A movement that calls for a Jewish homeland in Palestine

_____ **6.** Another name for the Asian part of Turkey

_____ **7.** A narrow body of water that lies between Europe and Asia

_____ **8.** Capital of Syria and one of the oldest cities in the world

_____ **9.** A light metal used in aerospace industries

_____ **10.** Nonreligious

a. Damascus
b. mandates
c. Haifa
d. magnesium
e. Anatolia
f. Kurds
g. secular
h. Bosporus
i. Ankara
j. potash
k. Zionism
l. Beirut

FILL IN THE BLANK *(3 points each)* For each of the following statements, fill in the blank with the appropriate word, phrase, or place.

1. The ______________________ moderates the climate of the western coast of Turkey.

2. The island of Cyprus is divided between ______________________ and ______________________ factions.

3. The lowest land on Earth's surface is the shore of the ______________________.

4. Due to many faults, folds, and mountain building, ______________________ experiences severe earthquakes.

5. The three most common languages in the eastern Mediterranean are ______________________.

6. Atatürk moved the capital of Turkey from Constantinople to

______________________.

7. ______________________ is the most technologically advanced country in the eastern Mediterranean.

8. The ______________________ wants to establish a separate Palestinian country.

9. The Crusades began when the ______________________ threatened to take Constantinople.

10. ______________________ are open-air markets.

UNDERSTANDING IDEAS *(3 points each)* For each of the following, write the letter of the best choice in the space provided.

______ **1.** What two mountain systems run across Anatolia?
- **a.** the Pontic and Caucasus Mountains
- **b.** the Taurus and Caucasus Mountains
- **c.** the Pontic and Taurus Mountains
- **d.** the Taurus and Pyrenees

______ **2.** What two deserts are found in the Eastern Mediterranean?
- **a.** Syrian and Sahara
- **b.** Sahel and Syrian
- **c.** Syrian and Negev
- **d.** Negev and Sahel

______ **3.** Which of the following is not a problem in Syria?
- **a.** high unemployment
- **b.** low birthrate
- **c.** weak educational system
- **d.** outdated technology

______ **4.** What was the capital of the Byzantine Empire?
- **a.** Damascus
- **b.** Constantinople
- **c.** Jerusalem
- **d.** Beirut

______ **5.** The regions of land that are disputed in Israel are
- **a.** the Gaza Strip, the West Bank, and the Golan Heights.
- **b.** Syria, Palestine, and the Gaza Strip.
- **c.** the West Bank, the Golan Heights, and Haifa.
- **d.** the Golan Heights, the Gaza Strip, and Tel Aviv.

PRACTICING SKILLS *(5 points each)* Study the diagram and answer the questions that follow.

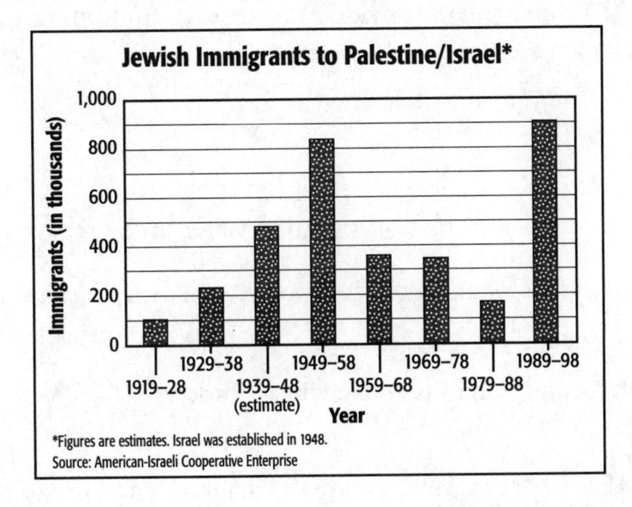

1. In what period did the greatest number of people immigrate to Israel?

____

____

2. To what can you attribute the large increase in immigration for the 1949–1958 period?

____

____

COMPOSING AN ESSAY *(15 points)* On a separate sheet of paper, write an essay in response to *one* of the following.

1. Summarize the political history of Israel since 1947.

2. How have the region's landforms and resources affected its economic growth?

Name ______________________ Class ____________ Date ____________

Chapter Test Form B

The Eastern Mediterranean

SHORT ANSWER *(10 points each)* Provide brief answers for each of the following. Use examples to support your answers.

1. What are the major landforms of Turkey?

2. What is the PLO?

3. What are three important sources of irrigation water in the region?

4. Why is the control of water an important political issue in the eastern Mediterranean?

5. What are the ethnic and religious challenges in Turkey?

PRACTICING SKILLS *(10 points each)* Study the diagram and answer the questions that follow.

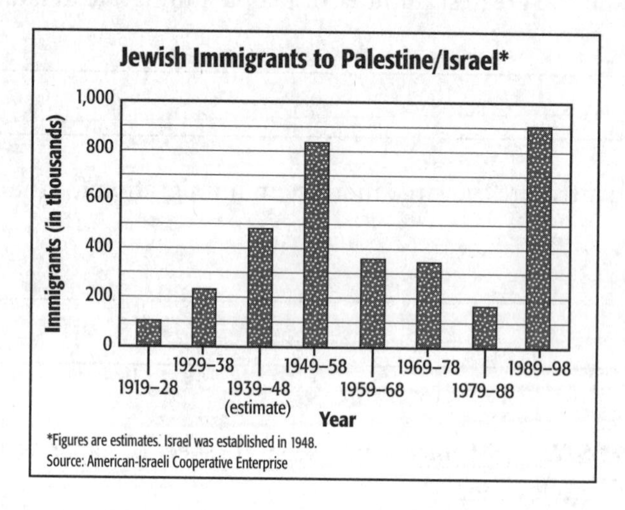

1. What period witnessed the largest jump in immigration numbers?

2. Why do you think the number of immigrants dropped between 1978 and 1979?

COMPOSING AN ESSAY *(30 points)* On a separate sheet of paper, write an essay in response to *one* of the following.

1. Why do many Israelis and Arabs view political issues in Palestine differently?

2. Explain how religion shapes the cultural patterns of the region.

3. Compare and contrast the economies of Israel and Jordan.

4. How does religion affect the political stability of the region?

Name ______________________ Class ____________ Date ____________

UNIT 6 Unit Test Form A

Southwest Asia

REVIEWING FACTS

MATCHING *(3 points each)* In the space provided, write the letter of the term or place that matches each description. Some answers will not be used.

_____ **1.** A line of hereditary rulers

_____ **2.** Nonreligious

_____ **3.** A narrow body of water that lies between Europe and Asia

_____ **4.** Shia religious leader of the highest order

_____ **5.** Government based on religious laws

_____ **6.** Another name for the Asian part of Turkey

_____ **7.** River that begins in a humid region and flows to a dry region

_____ **8.** A large port city in Israel

_____ **9.** Is becoming wider as the African and Arabian plates move apart

_____ **10.** Territories placed under another country's control

a. imam
b. Bosporus
c. dynasty
d. exotic river
e. secular
f. Damascus
g. ayatollah
h. Red Sea
i. mandates
j. theocracy
k. Haifa
l. Anatolia

FILL IN THE BLANK *(3 points each)* For each of the following statements, fill in the blank with the appropriate word, phrase, or place.

1. ______________________ is the most technologically advanced country in the eastern Mediterranean.

2. The ______________________ are found along Iran's western coast.

3. ______________________ is the capital of Afghanistan.

4. The three most common languages in the eastern Mediterranean are ______________________.

5. Atatürk moved the capital of Turkey from Constantinople to ______________________.

6. The most common language in Iran is ______________________.

7. The island of Cyprus is divided between ______________________ and ______________________ factions.

8. The only wet climates in the Persian Gulf and Interior are produced by the ______________________.

9. The ______________________ moderates the climate of the western coast of Turkey.

10. The ______________________ developed what became known as the golden age of Persian culture.

UNDERSTANDING IDEAS *(3 points each)* For each of the following, write the letter of the best choice in the space provided.

______ **1.** Which country is not located on the Arabian Peninsula?
- **a.** Qatar
- **b.** Kuwait
- **c.** Iraq
- **d.** United Arab Emirates

______ **2.** Which of the following is not a problem in Syria?
- **a.** high unemployment
- **b.** low birthrate
- **c.** weak educational system
- **d.** outdated technology

______ **3.** What lies to the south of the Arabian Peninsula?
- **a.** Mediterranean Sea and Shatt al Arab
- **b.** Arabian Sea and Red Sea
- **c.** Persian Gulf and Shatt al Arab
- **d.** Arabian Sea and the Gulf of Aden

______ **4.** Which is not a basic theme in the Persian Gulf and Interior?
- **a.** oil
- **b.** democratic rule
- **c.** the role of Islam
- **d.** preservation of traditional leaders

______ **5.** The regions of the land that are disputed in Israel are
- **a.** the Gaza Strip, the West Bank, and the Golan Heights.
- **b.** Syria, Palestine, and the Gaza Strip.
- **c.** the West Bank, the Golan Heights, and Haifa.
- **d.** the Golan Heights, the Gaza Strip, and Tel Aviv.

Name ____________________ Class ______________ Date ______________

PRACTICING SKILLS *(5 points each)* Study the graph and answer the questions that follow.

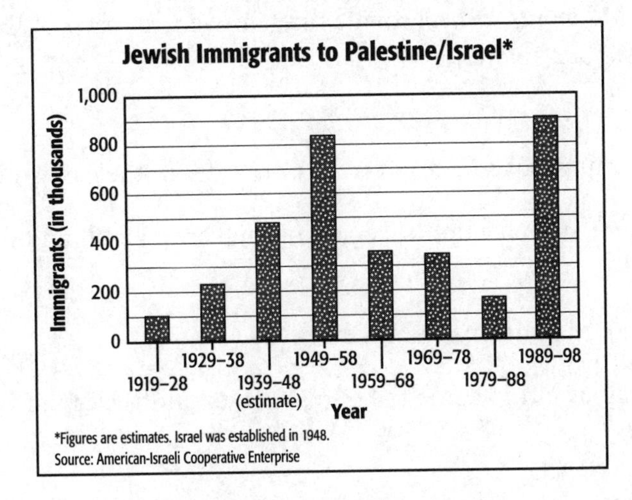

1. In what period did the greatest number of people immigrate to Israel?

____

2. To what can you attribute the large increase in immigration for the 1939–1958 period?

____

COMPOSING AN ESSAY *(15 points)* On a separate sheet of paper, write an essay in response to *one* of the following.

1. Write a brief history of the Persian Gulf region.

2. How have the landforms and resources of the region affected the economic growth of the eastern Mediterranean?

Name ______________________ Class ______________ Date ______________

Unit Test Form B

Southwest Asia

SHORT ANSWER *(10 points each)* Provide brief answers for each of the following. Use examples to support your answers.

1. What is the PLO?

2. What are three important sources of irrigation water in the eastern Mediterranean?

3. List the countries that make up the region of the Persian Gulf and interior Southwest Asia.

4. What are the major landforms of Turkey?

5. How did the Arabic culture come to be so prevalent in the Persian Gulf and interior Southwest Asia?

PRACTICING SKILLS *(10 points each)* Study the graph and answer the questions that follow.

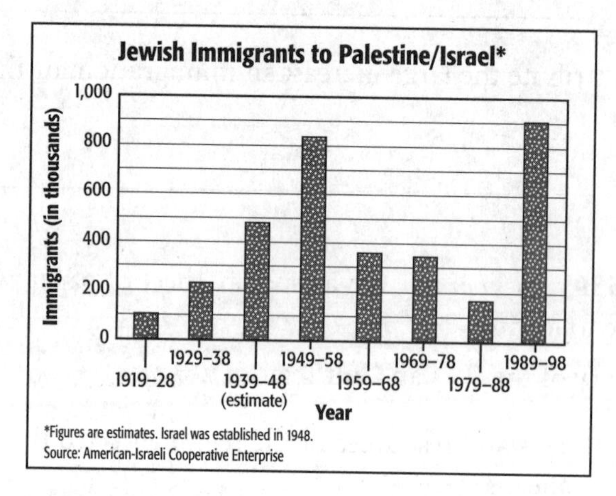

1. What period witnessed the largest jump in immigration numbers?

2. Why do you think the number of immigrants dropped between 1978 and 1979?

COMPOSING AN ESSAY *(30 points)* On a separate sheet of paper, write an essay in response to *one* of the following.

1. Why is the political stability of the Persian Gulf important to countries around the world?

2. Compare and contrast the economies of Israel and Jordan.

3. Explain how religion shapes the cultural patterns of the eastern Mediterranean.

4. How does the availability of water in the Persian Gulf and interior Southwest Asia shape economic activity?

Name ______________________ Class ______________ Date ____________

Chapter Test Form A

North Africa

REVIEWING FACTS

MATCHING *(3 points each)* In the space provided, write the letter of the term or place that matches each description. Some answers will not be used.

__i__ **1.** Peasant farmers

__g__ **2.** Gravel-covered plain

__h__ **3.** Pictures and symbols used in an Egyptian writing system

__b__ **4.** Capital city of Libya

__j__ **5.** Moroccan city that includes a famous medina

__f__ **6.** Ancient Egyptian rulers

__k__ **7.** Dry streambed

__a__ **8.** A sea of sand in the desert

__c__ **9.** Fertile finely ground soil

__l__ **10.** Capital of Tunisia

a. erg
b. Tripoli
c. silt
d. depressions
e. free port
f. pharaohs
g. reg
h. hieroglyphs
i. fellahin
j. Fès
k. wadis
l. Tunis

FILL IN THE BLANK *(3 points each)* For each of the following statements, fill in the blank with the appropriate word, phrase, or place.

1. Morocco is the world's largest exporter of ______________________.

2. ______________________ is the largest urban area in North Africa.

3. North Africa is bordered on the east by the ______________________ Sea.

4. The completion of the Aswān High Dam has led to a decrease in the amount of ______________________ deposited by the Nile River.

5. ______________________ took control of Egypt in 1882 to gain control of the Suez Canal.

6. A major challenge facing urban areas of North Africa is ______________________.

7. The ______________________ is a hot dry wind that sweeps across the Sahara.

8. ______________________ is a free port in Morocco.

9. Due to its close proximity to Europe, ______________________ has a great diversity of plant and animal life.

10. Of the countries of North Africa, only ______________________ does not have a strong farming sector.

UNDERSTANDING IDEAS *(3 points each)* For each of the following, write the letter of the best choice in the space provided.

__d__ **1.** Where did the Phoenician civilization originate?
- **a.** Libya
- **b.** the Sinai Peninsula
- **c.** Carthage
- **d.** Lebanon

__a__ **2.** In what location are few taxes placed on goods from other countries?
- **a.** free port
- **b.** fellahin
- **c.** wadis
- **d.** command economy

__a__ **3.** The Nile River empties into what body of water?
- **a.** Mediterranean Sea
- **b.** Red Sea
- **c.** Suez Canal
- **d.** Atlantic Ocean

__b__ **4.** Desert nomads living along the Sinai Peninsula in Egypt are called
- **a.** Berbers.
- **b.** Bedouins.
- **c.** wadis.
- **d.** Moors.

__c__ **5.** What has severely affected tourism in Algeria?
- **a.** drought
- **b.** inflation
- **c.** political unrest
- **d.** overcrowding

PRACTICING SKILLS *(5 points each)* Study the diagram and answer the questions that follow.

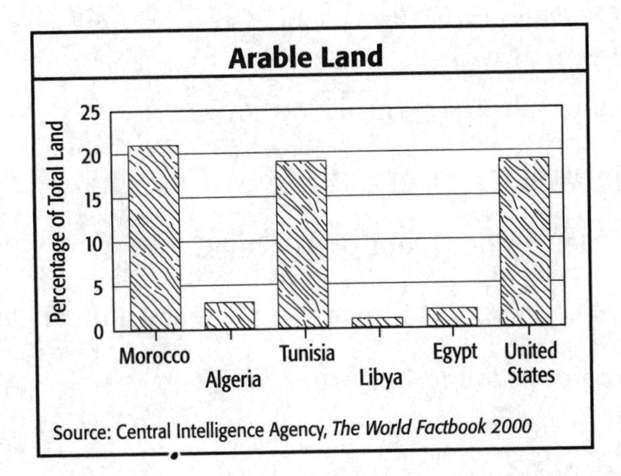

1. Which country has a percentage of arable land equal to that of the United States?

____

2. Which country has the lowest percentage of arable land?

____

COMPOSING AN ESSAY *(15 points)* On a separate sheet of paper, write an essay in response to *one* of the following.

1. Contrast the positions of Islamic fundamentalists and the region's governments on the proper role of Islam in government.

2. Discuss the benefits and problems created by the Aswān High Dam.

Chapter Test Form B

North Africa

SHORT ANSWER *(10 points each)* Provide brief answers for each of the following. Use examples to support your answers.

1. Name the civilizations that have ruled North Africa.
2. What are the capitals of the region's countries?
3. Why did Great Britain want to control the Suez Canal?
4. Why does Morocco have a great diversity of plant and animal life?
5. List the major landforms found in North Africa.

PRACTICING SKILLS *(10 points each)* Study the diagram and answer the questions that follow.

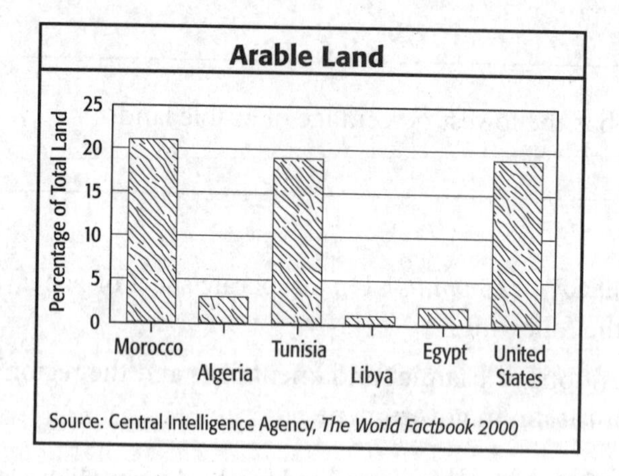

1. Which two North African countries have the highest percentages of arable land?
2. Why might these countries have a higher percentage of arable land than other countries in the region?

COMPOSING AN ESSAY *(30 points)* On a separate sheet of paper, write an essay in response to *one* of the following.

1. Why might migration out of North Africa present a challenge to the economies of the region?
2. What are the buildings and streets of medinas like? Why are they like this?
3. What are the natural resources of North Africa?
4. How has rapid population increase caused difficulties for the economies of North Africa?

Name ______________________ Class ______________ Date ______________

Chapter Test Form A

West and Central Africa

REVIEWING FACTS

MATCHING *(3 points each)* In the space provided, write the letter of the term or place that matches each description. Some answers will not be used.

l 1. Capital city of Democratic Republic of Congo
g 2. Main food crop of a region
b 3. Largest city in Nigeria
k 4. Region of semiarid climate along southern edge of Sahara
i 5. System under which a country produces some products exclusively for export and some for local use
a 6. Spread of desert conditions
f 7. River that flows northward from Zambia
j 8. River that empties into the Gulf of Guinea
e 9. Important center of trade during the Ghana Empire
d 10. Drought-resistant grain

a. desertification
b. Lagos
c. Accra
d. sorghum
e. Tombouctou
f. Congo River
g. staple
h. Abidjan
i. dual economy
j. Niger River
k. Sahel
l. Kinshasa

FILL IN THE BLANK *(3 points each)* For each of the following statements, fill in the blank with the appropriate word, phrase, or place.

1. ______________ and ______________ are major cash crops grown in West Africa.

2. ______________ is the main obstacle to education in the region.

3. Most of the region's countries export ______________ goods.

4. West and Central Africa have a ______________ climate.

5. Gabon is one of the richest countries in Africa because of its ______________.

6. The ______________ Empire was the earliest powerful kingdom in the region.

7. Landlocked ______________________ is one of the poorest countries in the world.

8. ______________________ is a desert region in eastern Mauritania.

9. The cacao tree is native to ______________________.

10. The development of a ______________________ economy has seriously affected traditional farming.

UNDERSTANDING IDEAS *(3 points each)* For each of the following, write the letter of the best choice in the space provided.

__d__ **1.** All of the following are challenges facing the countries of West and Central Africa except
- **a.** disease
- **b.** overpopulation
- **c.** lack of economic development
- **d.** aging populations

__a__ **2.** Which of the following has contributed to the desertification of the Sahel?
- **a.** growing population
- **b.** excessive winds
- **c.** heavy rains
- **d.** plantation farming

__d__ **3.** Why did Europeans originally explore the coast of West Africa?
- **a.** They were looking for oil.
- **b.** They were looking for slave labor.
- **c.** They were looking for spices.
- **d.** They were looking for a water route to Asia.

__c__ **4.** The two major rivers of West and Central Africa are
- **a.** the Congo and Nile Rivers.
- **b.** the White and Blue Nile Rivers.
- **c.** the Niger and Congo Rivers.
- **d.** the Niger and Nile Rivers.

__c__ **5.** Crops grown in West and Central Africa include all of the following except
- **a.** cacao
- **b.** peanuts
- **c.** corn
- **d.** cassava

Name ______________________ Class ______________ Date ____________

PRACTICING SKILLS *(5 points each)* Study the diagram and answer the questions that follow.

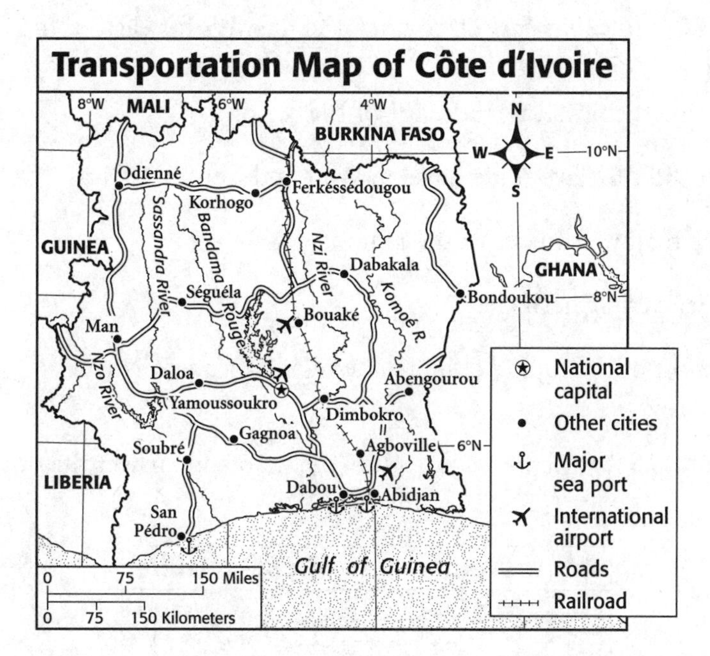

1. Through which city would a driver pass on a direct trip from Odienné to Ferkéssédougou?

____

2. Which city is served by a railroad line, an airport, and a sea port?

____

COMPOSING AN ESSAY *(15 points)* On a separate sheet of paper, write an essay in response to *one* of the following.

1. Discuss the steps that could be taken to lessen desertification in the region.

2. Choose a region in West and Central Africa and write a description of the climate zone in which it lies and the resulting weather for that region.

Name ______________________ Class ______________ Date ____________

Chapter Test Form B

West and Central Africa

SHORT ANSWER *(10 points each)* Provide brief answers for each of the following. Use examples to support your answers.

1. Briefly describe the physical landscape of the region.

2. What effects did the slave trade have on West and Central Africa?

3. What are the major resources of the region?

4. What is Africa's triple heritage?

5. Where are most of the capital cities of the region located? Why?

PRACTICING SKILLS *(10 points each)* Study the map and answer the questions that follow.

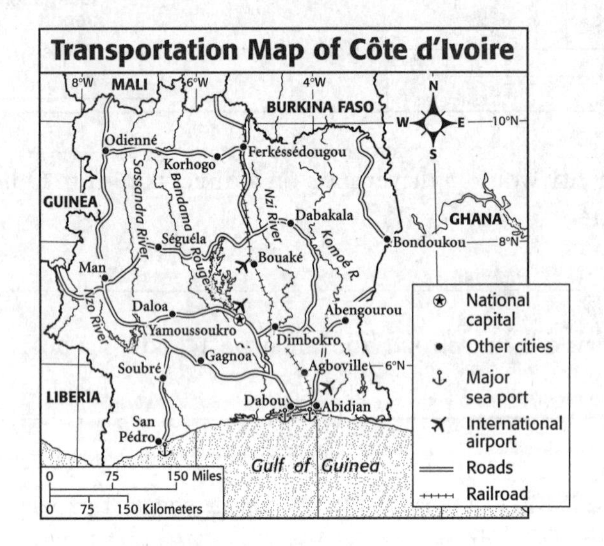

1. Why might Bondoukou be difficult to reach from within Côte d'Ivoire?

2. Which city is located near the head of the Nzo River?

COMPOSING AN ESSAY *(30 points)* On a separate sheet of paper, write an essay in response to *one* of the following.

1. Explain why literacy rates are so low in the region.

2. What reasons have been put forth to explain the unusual courses of the Congo and Niger Rivers?

3. Describe family life in the region.

4. Discuss the effects of European colonization on the region.

Name ______________________ Class ______________ Date ____________

Chapter Test Form A

East Africa

REVIEWING FACTS

MATCHING *(3 points each)* In the space provided, write the letter of the term or place that matches each description. Some answers will not be used.

d 1. Africa's largest lake

a 2. Earliest human remains have been found here

i 3. Cream-colored material that makes up the tusks of elephants

l 4. Sap from acacia trees

f 5. Major port located in Tanzania

b 6. The Western Rift Valley begins near here

k 7. Durable plant fiber used to make rope

e 8. Capital of Sudan

h 9. Carries the human disease known as sleeping sickness

j 10. The intentional destruction of a people

a. Olduvai Gorge
b. Lake Malawi
c. Serengeti Plain
d. Lake Victoria
e. Khartoum
f. Dar es Salaam
g. Swahili
h. tsetse fly
i. ivory
j. genocide
k. sisal
l. gum arabic

FILL IN THE BLANK *(3 points each)* For each of the following statements, fill in the blank with the appropriate word, phrase, or place.

1. One of the earliest powerful kingdoms of East Africa was ______________________.

2. ______________________ has the highest per capita GDP in the region.

3. The last East African colony to win independence was ______________________.

4. One of the major problems of the region is a rapidly growing ______________________.

5. The Blue Nile and the White Nile join in ______________________.

6. ______________________ is a staple grain in East Africa.

7. Most of the waters of the Nile River are provided by the ______________________ Nile.

8. ______________________ is the region's largest city.

9. Linguistically, the Masai and Tutsi are ______________________ peoples.

10. The climates found in East Africa vary greatly due to ______________________ and latitude.

UNDERSTANDING IDEAS *(3 points each)* For each of the following, write the letter of the best choice in the space provided.

__a__ **1.** Where does the Blue Nile begin?
- **a.** in the highlands of northern Ethiopia
- **b.** in Lake Victoria
- **c.** in Lake Albert
- **d.** in wetlands called the Sudd

__a__ **2.** The Kikuyu and Hutu are both from which group?
- **a.** Bantu speakers
- **b.** Cushitic speakers
- **c.** Nilotic peoples
- **d.** Tutsi peoples

__a__ **3.** Which country is changing from a command economy to a market economy?
- **a.** Ethiopia
- **b.** Somalia
- **c.** Eritrea
- **d.** Rwanda

__b__ **4.** The Eastern Rift Valley begins in
- **a.** Tanzania.
- **b.** Mozambique.
- **c.** Kenya.
- **d.** Sudan.

__b__ **5.** Which early culture built pyramids?
- **a.** Aksum
- **b.** Kush
- **c.** Swahili
- **d.** Hutu

PRACTICING SKILLS *(5 points each)* Study the table and answer the questions that follow.

Religions of East Africa

Country	Christian	Muslim	Traditional Beliefs	Other
Burundi	60%	1%	39%	
Djibouti		94%		6%
Eritrea	45%	45%		10%
Ethiopia	37%	43%	17%	3%
Kenya	66%	6%	26%	2%
Rwanda	45%		50%	5%
Somalia		100%		
Sudan	5%	70%	20%	5%
Tanzania*	26%	31%	42%	1%
Uganda	66%	16%	18%	

*except Zanzibar, which is 99% Muslim

Source: *The DK Geography of the World*

1. Which country has the most even distribution of Christianity, Islam, and traditional beliefs?

____

2. Which two countries have the largest percentage of Christian citizens?

____

COMPOSING AN ESSAY *(15 points)* On a separate sheet of paper, write an essay in response to *one* of the following.

1. Why can it be said that the wildlife of the Serengeti Plain owes its survival in part to the tsetse fly?

2. Discuss how European colonization both helped and harmed East Africa.

Name ______________________ Class ______________ Date ____________

Chapter Test Form B

East Africa

SHORT ANSWER *(10 points each)* Provide brief answers for each of the following. Use examples to support your answers.

1. Name the three linguistic groups into which the ethnic groups of East Africa can be divided.

2. How were the rift valleys of East Africa created?

3. What political and economic tensions have added to the problems facing East Africa?

4. Name six of the region's agricultural products.

5. Describe the roles of East African men and women in agriculture.

PRACTICING SKILLS *(10 points each)* Study the table and answer the questions that follow.

Religions of East Africa

Country	Christian	Muslim	Traditional Beliefs	Other
Burundi	60%	1%	39%	
Djibouti		94%		6%
Eritrea	45%	45%		10%
Ethiopia	37%	43%	17%	3%
Kenya	66%	6%	26%	2%
Rwanda	45%		50%	5%
Somalia		100%		
Sudan	5%	70%	20%	5%
Tanzania*	26%	31%	42%	1%
Uganda	66%	16%	18%	

*except Zanzibar, which is 99% Muslim

Source: *The DK Geography of the World*

1. In which country does the largest percentage of the population adhere to traditional beliefs?

2. Which country is the most homogeneous in terms of religion?

COMPOSING AN ESSAY *(30 points)* On a separate sheet of paper, write an essay in response to *one* of the following.

1. Summarize the steps taken by the government of Ethiopia to create a command economy in the 1970s.

2. What did European explorers of the 1800s discover about the source of the Nile?

3. How did the language known as Swahili develop?

4. Discuss the rapid growth of East African cities and the effects of that growth.

Name ______________________ Class ______________ Date ____________

Chapter Test Form A

Southern Africa

REVIEWING FACTS

MATCHING *(3 points each)* In the space provided, write the letter of the term or place that matches each description. Some answers will not be used.

__b__ **1.** Largest township in South Africa

__l__ **2.** A system of laws that enforced "separateness"

__e__ **3.** City that is government center of Zimbabwe

__i__ **4.** People who do not work for a formal business

__k__ **5.** An exclave of Angola

__g__ **6.** A city in Mozambique founded by the Portuguese

__h__ **7.** A steep face at the edge of a plateau

__a__ **8.** Many different types of plants and animals

__d__ **9.** Penalties intended to force a country to change its policies

__c__ **10.** Grassland

a. biodiversity
b. Soweto
c. veld
d. sanctions
e. Harare
f. Afrikaners
g. Maputo
h. escarpment
i. informal sector
j. Drakensberg
k. Cabinda
l. apartheid

FILL IN THE BLANK *(3 points each)* For each of the following statements, fill in the blank with the appropriate word, phrase, or place.

1. The major mountain range of southern Africa is the ______________________.

2. The Okavango Swamps are located in northern ______________________.

3. ______________________ has the most developed economy in Africa.

4. The first Europeans to explore southern Africa were ______________________.

5. The first black president of South Africa was ______________________.

6. ______________________ is the most common economic activity in southern Africa.

7. Popular tourist attractions in southern Africa include its

______________________ and ______________________.

8. The ______________________ Church is the largest Christian denomination in South Africa.

Name ______________________ Class ______________ Date ____________

9. Zambia is a major producer of ______________________.

10. The ______________________ River is the only river in southern Africa to drain into the Atlantic Ocean.

UNDERSTANDING IDEAS *(3 points each)* For each of the following, write the letter of the best choice in the space provided.

__d__ **1.** What skill did the early Bantu settlers bring to southern Africa?
- **a.** how to cultivate land
- **b.** how to irrigate farmland
- **c.** how to herd sheep
- **d.** how to make iron tools

__c__ **2.** Water in the Namib Desert comes from
- **a.** artesian wells.
- **b.** the Limpopo River.
- **c.** dew and fog.
- **d.** monsoon weather.

__d__ **3.** What was the cause of the Boer War?
- **a.** conflicts between Boers and Bantu
- **b.** conflicts over water rights
- **c.** disputes over land ownership
- **d.** desire to control the region's mineral wealth

__d__ **4.** Why is life expectancy declining in southern Africa?
- **a.** poverty
- **b.** pollution
- **c.** declining birthrate
- **d.** high rate of HIV/AIDS infection

__b__ **5.** Which country does not have productive diamond mines?
- **a.** South Africa
- **b.** Zambia
- **c.** Botswana
- **d.** Namibia

PRACTICING SKILLS *(5 points each)* Study the map and answer the questions that follow.

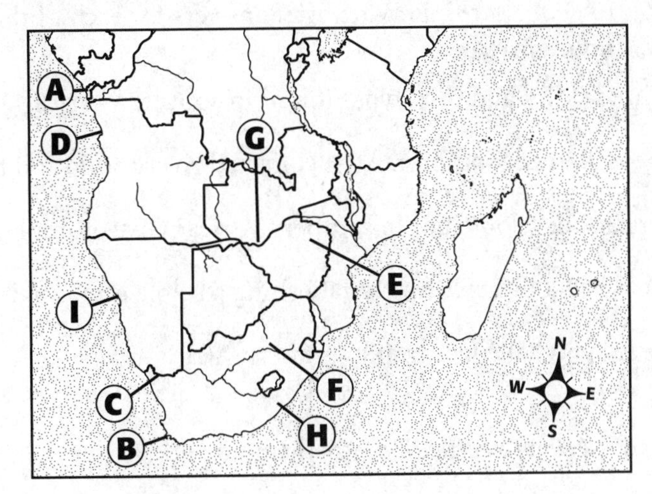

1. What city is marked by the letter E?

____

2. What area is marked by the letter A?

____

COMPOSING AN ESSAY *(15 points)* On a separate sheet of paper, write an essay in response to *one* of the following.

1. Now that apartheid has ended in South Africa, how is the country progressing economically and socially?

2. Summarize the history of the Dutch in South Africa.

Name ______________________ Class ______________ Date ____________

Chapter Test Form B

Southern Africa

SHORT ANSWER *(10 points each)* Provide brief answers for each of the following. Use examples to support your answers.

1. Describe the three main landform types found in southern Africa.
2. How has the geography of southern Africa affected river traffic and trade?
3. How did Europeans develop cities in southern Africa? Give examples.
4. Why has Zambia had a difficult time establishing a stable economy?
5. What are three major challenges facing southern Africa today?

PRACTICING SKILLS *(10 points each)* Study the map and answer the questions that follow.

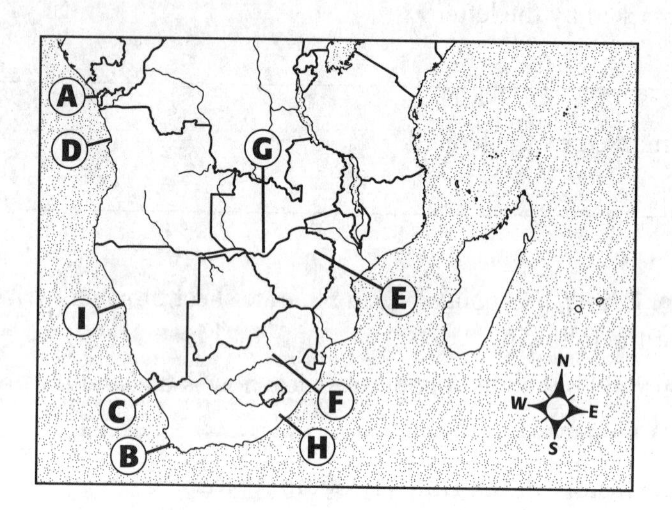

1. What area is marked by the letter I?
2. What city is marked by the letter F?

COMPOSING AN ESSAY *(30 points)* On a separate sheet of paper, write an essay in response to *one* of the following.

1. Compare and contrast market-oriented and subsistence agriculture.
2. Why is the Namib Desert especially dry? What kind of life is able to exist there?
3. Why do many African countries still use European languages? Give examples.
4. Describe the economic history of Botswana.

Name ______________________ Class ____________ Date __________

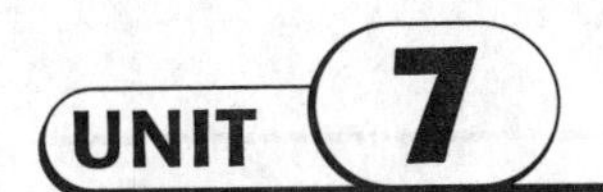

Unit Test Form A

Africa

REVIEWING FACTS

MATCHING *(3 points each)* In the space provided, write the letter of the term or place that matches each description. Some answers will not be used.

______ **1.** Important center of trade during the Ghana Empire

______ **2.** The intentional destruction of a people

______ **3.** Dry streambeds

______ **4.** A steep face at the edge of a plateau

______ **5.** Capital city of Libya

______ **6.** Region of semiarid climate along southern edge of Sahara

______ **7.** Grassland

______ **8.** An exclave of Angola

______ **9.** Sap from acacia trees

______ **10.** Main food crop of a region

a. Tripoli
b. tsetse fly
c. Cabinda
d. Tombouctou
e. gum arabic
f. wadis
g. staple
h. veld
i. genocide
j. fellahin
k. escarpment
l. Sahel

FILL IN THE BLANK *(3 points each)* For each of the following statements, fill in the blank with the appropriate word, phrase, or place.

1. The development of a ______________________ economy has seriously affected traditional farming in West and Central Africa.

2. ______________________ took control of Egypt in 1882 to gain control of the Suez Canal.

3. One of the earliest powerful kingdoms of East Africa was ______________________.

4. Gabon is one of the richest countries in Africa because of its ______________________.

5. Linguistically, the Masai and the Tutsi are ______________________ peoples.

6. North Africa is bordered on the east by the ______________________ Sea.

7. Due to its close proximity to Europe, ______________________ has a great diversity of plant and animal life.

8. West and Central Africa have a ______________________ climate.

9. The ______________________ River is the only river in southern Africa to drain into the Atlantic Ocean.

10. ______________________ is a staple grain in East Africa.

UNDERSTANDING IDEAS *(3 points each)* For each of the following, write the letter of the best choice in the space provided.

______ **1.** The Eastern Rift Valley begins in
- **a.** Tanzania.
- **b.** Mozambique.
- **c.** Kenya.
- **d.** Sudan.

______ **2.** Which early culture built pyramids?
- **a.** Aksum
- **b.** Kush
- **c.** Swahili
- **d.** Hutu

______ **3.** What skill did the early Bantu settlers bring to southern Africa?
- **a.** how to cultivate land
- **b.** how to irrigate farmland
- **c.** how to herd sheep
- **d.** how to make iron tools

______ **4.** Which of the following has contributed to the desertification of the Sahel?
- **a.** growing population
- **b.** excessive winds
- **c.** heavy rains
- **d.** plantation farming

______ **5.** Desert nomads living along the Sinai Peninsula in Egypt are called
- **a.** Berbers.
- **b.** Bedouins.
- **c.** wadis
- **d.** Moors.

Name ______________________ Class ______________ Date ____________

PRACTICING SKILLS *(5 points each)* Study the map and answer the questions that follow.

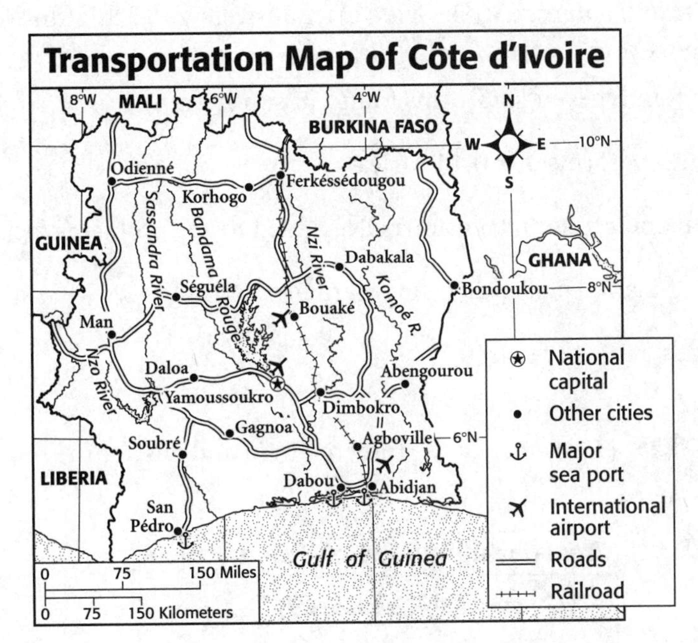

1. Through which city would a driver pass on a direct trip from Odienné to Ferkéssédougou?

____

2. Which city is served by a railroad line, an airport, and a sea port?

____

COMPOSING AN ESSAY *(15 points)* On a separate sheet of paper, write an essay in response to *one* of the following.

1. Now that apartheid has ended in South Africa, how is the country progressing economically and socially?

2. Discuss the benefits and problems created by the Aswān High Dam.

Name ________________ Class ________ Date ________

Unit Test Form B

Africa

SHORT ANSWER *(10 points each)* Provide brief answers for each of the following. Use examples to support your answers.

1. Name three main types of landforms found in southern Africa.

2. List the major landforms of North Africa.

3. What political and economic tensions have added to the problems facing East Africa?

4. What are three major challenges facing southern Africa today?

5. What is Africa's triple heritage?

PRACTICING SKILLS *(10 points each)* Study the map and answer the questions that follow.

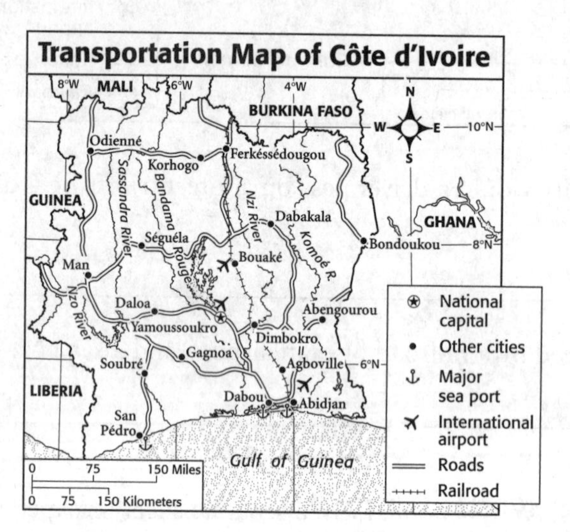

1. Why might Bondoukou be difficult to reach from within Côte d'Ivoire?

2. Which city is located near the head of the Nzo River?

COMPOSING AN ESSAY *(30 points)* On a separate sheet of paper, write an essay in response to *one* of the following.

1. Compare and contrast market-oriented and subsistence agriculture.

2. How has rapid population increase caused difficulties for the economies of North Africa?

3. What reasons have been put forth to explain the unusual courses of the Congo and Niger Rivers?

4. How did the language known as Swahili develop?

Name ______________________ Class ______________ Date ______________

Chapter Test Form A

India

REVIEWING FACTS

MATCHING *(3 points each)* In the space provided, write the letter of the term or place that matches each description. Some answers will not be used.

______ **1.** Large city that is center of Indian movie industry

______ **2.** Indian troops under British command

______ **3.** All the gods of a religion

______ **4.** Small-scale industries run out of the home

______ **5.** One of India's major landform regions

______ **6.** Plant fiber used to make burlap and rope

______ **7.** A language from which modern-day Hindi is derived

______ **8.** Large landmass that is smaller than a continent

______ **9.** River that flows through the Deccan Plateau

______ **10.** City that is the center of India's computer industry

a. Sanskrit
b. Mumbai
c. Brahmaputra River
d. Bangalore
e. sepoys
f. cottage industries
g. pantheon
h. subcontinent
i. jute
j. Deccan Plateau
k. partition
l. Narmada River

FILL IN THE BLANK *(3 points each)* For each of the following statements, fill in the blank with the appropriate word, phrase, or place.

1. ______________________ is the national language of India.

2. India is separated from the rest of Asia by the ______________________.

3. The earliest known civilization in India was the ______________________ civilization.

4. More than 70 percent of the Indian population lives in ______________________.

5. The ______________________ River eventually joins the Ganges in Bangladesh.

6. The first Europeans to arrive in India were the ______________________.

7. India's ______________________ needs to be strengthened to attract foreign investment.

8. India must import ______________________ to meet its energy needs.

9. The basis of India's economy is ______________________.

10. The capital of India is ______________________.

UNDERSTANDING IDEAS *(3 points each)* For each of the following, write the letter of the best choice in the space provided.

______ **1.** Which of the following is not a crop of which India produces more than any other country?
a. peanuts
b. sesame
c. rice
d. tea

______ **2.** Which religion is defined by a strict moral code based on preserving life?
a. Jainism
b. Sikhism
c. Buddhism
d. Hinduism

______ **3.** Which river does not flow through the Deccan Plateau?
a. Narmada
b. Ganges
c. Godavari
d. Krishna

______ **4.** Which city is a holy city for Hindus?
a. Bangalore
b. Kolkata
c. Chennai
d. Varanasi

______ **5.** Who reunited the Mughal Empire and expanded it into central India?
a. Akbar
b. Bābur
c. Shāh Jāhan
d. Aurangzeb

PRACTICING SKILLS *(5 points each)* Study the graph and answer the questions that follow.

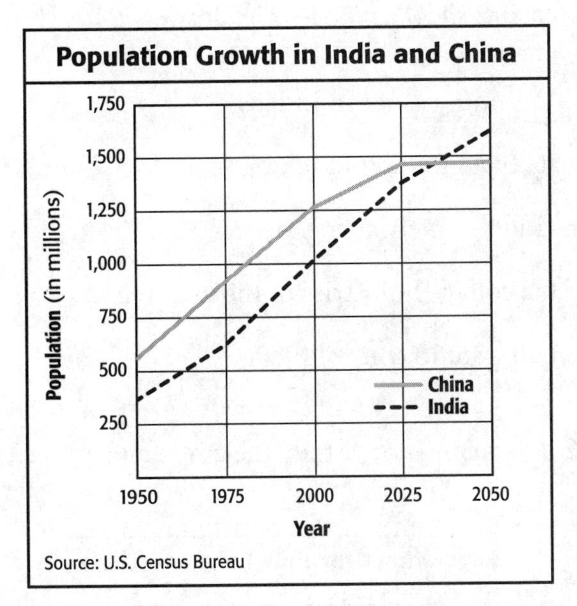

1. When is India expected to overtake China in population growth rate?

____

2. Describe the current and future population growth rates of the two countries.

____

COMPOSING AN ESSAY *(15 points)* On a separate sheet of paper, write an essay in response to *one* of the following.

1. What were the main elements of the Green Revolution? What was the final outcome of the program?

2. How did India achieve independence?

Name ______________________ Class ______________ Date ______________

Chapter Test Form B

India

SHORT ANSWER *(10 points each)* Provide brief answers for each of the following. Use examples to support your answers.

1. What creates the monsoon seasons in India?

2. Name the six climate types found in India.

3. Why was India partitioned after independence?

4. What are some of the cottage industries in India?

5. Why did Great Britain want to control India?

PRACTICING SKILLS *(10 points each)* Study the graph and answer the questions that follow.

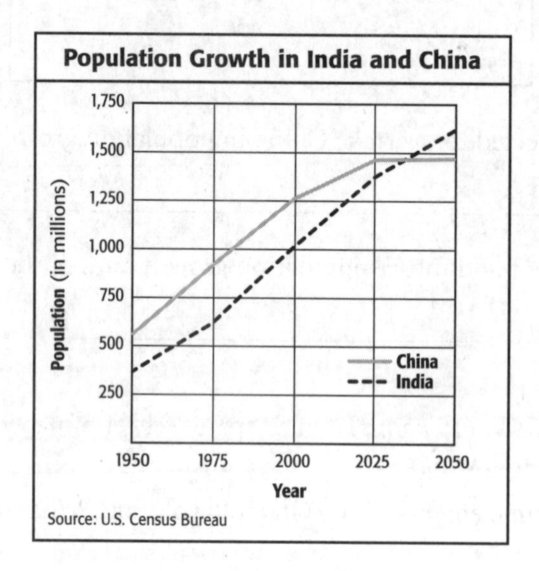

1. What is China's population expected to be in 2050?

2. Which country has a smaller population today?

COMPOSING AN ESSAY *(30 points)* On a separate sheet of paper, write an essay in response to *one* of the following.

1. Describe the plant and animal life of India. Give examples of typical animals.

2. Why have India's rivers and soils been called its most valuable resource? Give specific information to support your answer.

3. Briefly discuss the major events and empires of India's early history to 1700.

4. Why did the sepoys revolt against the British?

Name ______________________ Class ______________ Date ____________

Chapter Test Form A

The Indian Perimeter

REVIEWING FACTS

MATCHING *(3 points each)* In the space provided, write the letter that matches each description. Choose your answers from the list below. Some answers will not be used.

_____ **1.** The former name of Sri Lanka

_____ **2.** The official language of Pakistan

_____ **3.** Pakistan's main farming region

_____ **4.** A strip of Gangetic Plain that contains Nepal's main farming region

_____ **5.** A form of carbon used in pencil lead

_____ **6.** Largest city of the Harappan civilization

_____ **7.** The largest city in Pakistan

_____ **8.** Large waves of ocean water that wash ashore like a very high tide

_____ **9.** Capital of Bhutan

_____ **10.** A country that allows another stronger country to provide protection

a. Indus Valley
b. Mohenjo Daro
c. graphite
d. Tarai
e. storm surge
f. Ceylon
g. Bengali
h. Islamabad
i. protectorate
j. Urdu
k. Karachi
l. Thimphu

FILL IN THE BLANK *(3 points each)* For each of the following statements, fill in the blank with the appropriate word, phrase, or place.

1. ______________________ sells electricity to India.

2. The largest ethnic group in ______________________ is the Bhote.

3. In terms of religion, the majority of Pakistanis are followers of

______________________.

4. Bangladesh was formerly known as ______________________.

5. The capital of Sri Lanka is ______________________.

6. Powerful tropical cyclones that occur in the western Pacific are known as

______________________.

7. The ______________________ is a major factor in the area's agriculture.

8. Literacy rates in the Indian Perimeter are generally ______________________.

9. ______________________ has been the scene of ethnic conflicts between the Tamil minority and the Sinhalese majority.

10. ______________________ and Bhutan are landlocked countries in the Himalayas.

UNDERSTANDING IDEAS *(3 points each)* For each of the following, write the letter of the best choice in the space provided.

______ **1.** This tropical country is made up of 1,200 small coral islands.
a. the Maldives
b. Sri Lanka
c. Andaman Islands
d. Bhutan

______ **2.** This empire flourished in Pakistan and Bangladesh from 1500–1700.
a. Dravidian
b. Mughal
c. Aryan
d. British

______ **3.** The topography of Bangladesh is continually reshaped by
a. earthquakes.
b. volcanoes.
c. floods.
d. erosion.

______ **4.** Which of the following is not a main religion in the Indian Perimeter?
a. Christianity
b. Buddhism
c. Islam
d. Hinduism

______ **5.** A significant factor in the slow economic development in the area is
a. political instability.
b. poor soil.
c. constant flooding.
d. a lack of natural resources.

Name ______________________ Class ______________ Date ____________

PRACTICING SKILLS *(5 points each)* Study the diagram and answer the questions that follow.

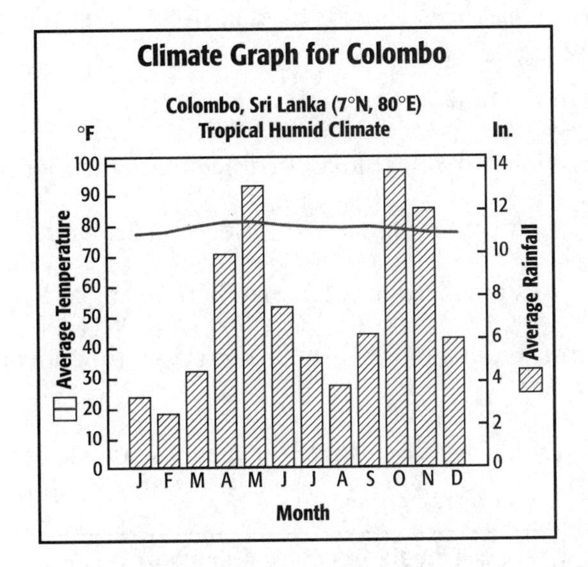

1. What pattern is shown by this graph? How might it affect agriculture?

____

____

2. What is the approximate average temperature for Colombo?

____

____

COMPOSING AN ESSAY *(15 points)* On a separate sheet of paper, write an essay in response to *one* of the following.

1. What factors have limited the economic growth of the Indian Perimeter?

2. Discuss the process by which Great Britain's decision to divide its Indian colony into India and Pakistan eventually led to the creation of Bangladesh.

Name ______________________ Class ____________ Date ____________

Chapter Test Form B

The Indian Perimeter

SHORT ANSWER *(10 points each)* Provide brief answers for each of the following. Use examples to support your answers.

1. List the countries of the Indian Perimeter.
2. Describe the main climates of the Indian Perimeter and their locations.
3. Describe the ethnic conflict that has led to violence in Sri Lanka.
4. List the major religions practiced in the Indian Perimeter.
5. Describe the bases of the economies of the countries in the Indian Perimeter.

PRACTICING SKILLS *(10 points each)* Study the diagram and answer the questions that follow.

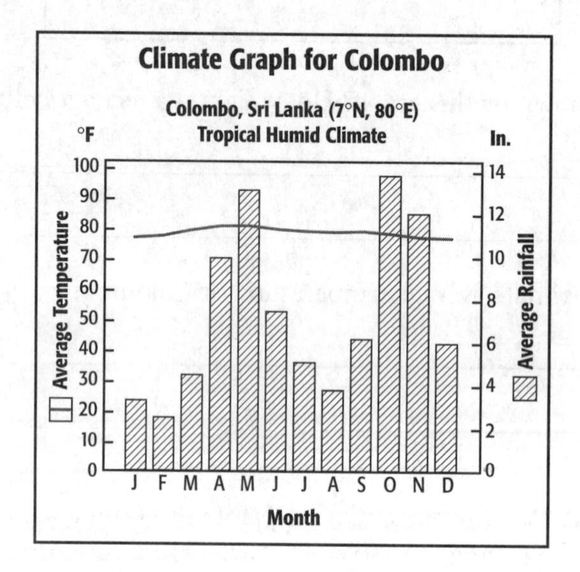

1. What month sees the least rainfall?
2. Compare and contrast the annual variation in temperature and rainfall.

COMPOSING AN ESSAY *(30 points)* On a separate sheet of paper, write an essay in response to *one* of the following.

1. Compare and contrast settlement patterns in Bangladesh and Sri Lanka.
2. Discuss Britain's decision to divide its Empire into the countries of India and Pakistan and the resulting problems that subdivided Pakistan.
3. Describe the factors that limit the economic growth of the Indian Perimeter.
4. How has Bangladesh's geographic location and climate impacted its development?

Name ______________________ Class ______________ Date ______________

Unit Test Form A

South Asia

REVIEWING FACTS

MATCHING *(3 points each)* In the space provided, write the letter of the term or place that matches each description. Some answers will not be used.

_____ **1.** River that flows through the Deccan Plateau

_____ **2.** A country that allows another, stronger country to provide protection

_____ **3.** Small-scale industries run out of a house

_____ **4.** A language from which modern-day Hindi is derived

_____ **5.** The former name of Sri Lanka

_____ **6.** A form of carbon used in pencil lead

_____ **7.** Indian troops under British command

_____ **8.** The official language of Pakistan

_____ **9.** All the gods of a religion

_____ **10.** A strip of Gangetic Plain that contains Nepal's main farming region

a. Narmada River
b. Karachi
c. Indus Valley
d. Sanskrit
e. Mumbai
f. Urdu
g. sepoys
h. graphite
i. Ceylon
j. cottage industries
k. pantheon
l. protectorate

FILL IN THE BLANK *(3 points each)* For each of the following statements, fill in the blank with the appropriate word, phrase, or place.

1. The first Europeans to arrive in India were ______________________.

2. Literacy rates in the Indian perimeter are generally ______________________.

3. In terms of religion, the majority of Pakistanis are followers of

______________________.

4. India is separated from the rest of Asia by the ______________________.

5. The capital of India is ______________________.

6. ______________________ and Bhutan are landlocked countries in the Himalayas.

7. The basis of India's economy is ______________________.

8. ______________________ has experienced ethnic conflicts between the Tamil minority and the Sinhalese majority.

9. The ______________________ River eventually joins the Ganges in Bangladesh.

10. Bangladesh was formerly known as ______________________.

UNDERSTANDING IDEAS *(3 points each)* For each of the following, write the letter of the best choice in the space provided.

_____ **1.** The topography of Bangladesh is continually reshaped by
- **a.** earthquakes.
- **b.** volcanoes.
- **c.** floods.
- **d.** erosion.

_____ **2.** Which religion is defined by a strict moral code based on preserving life?
- **a.** Jainism
- **b.** Sikhism
- **c.** Hinduism
- **d.** Buddhism

_____ **3.** A significant factor in the slow economic development of the Indian perimeter is
- **a.** political instability.
- **b.** poor soil.
- **c.** constant flooding.
- **d.** a lack of natural resources.

_____ **4.** Who reunited the Mughal Empire and expanded it into central India?
- **a.** Akbar
- **b.** Bābur
- **c.** Shāh Jāhan
- **d.** Aurangzeb

_____ **5.** Which of the following is not a crop of which India produces more than any other country?
- **a.** peanuts
- **b.** sesame
- **c.** rice
- **d.** tea

Name ______________________ Class ______________ Date ______________

PRACTICING SKILLS *(5 points each)* Study the graph and answer the questions that follow.

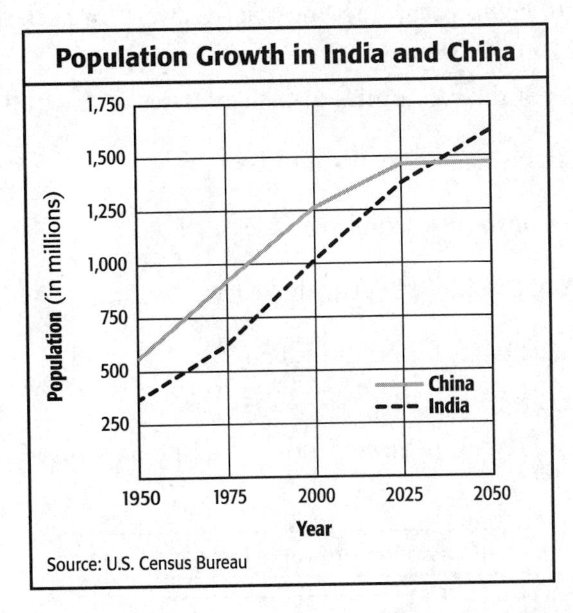

1. When is India expected to overtake China in population growth rate?

____

2. Describe the current and future population growth rates of the two countries.

____

COMPOSING AN ESSAY *(15 points)* On a separate sheet of paper, write an essay in response to *one* of the following.

1. How has its diversity of religions, languages, and cultures affected the economic growth of the Indian Perimeter?

2. What were the main elements of the Green Revolution? What was the final outcome of the program?

Name ______________________ Class ______________ Date ______________

Unit Test Form B

South Asia

SHORT ANSWER *(10 points each)* Provide brief answers for each of the following. Use examples to support your answers.

1. Describe the bases of the economies of the countries in the Indian Perimeter.

2. Why did Great Britain want to control India?

3. What creates the monsoon seasons in India?

4. List the countries of the Indian Perimeter.

5. Name the six climate types found in India.

PRACTICING SKILLS *(10 points each)* Study the graph and answer the questions that follow.

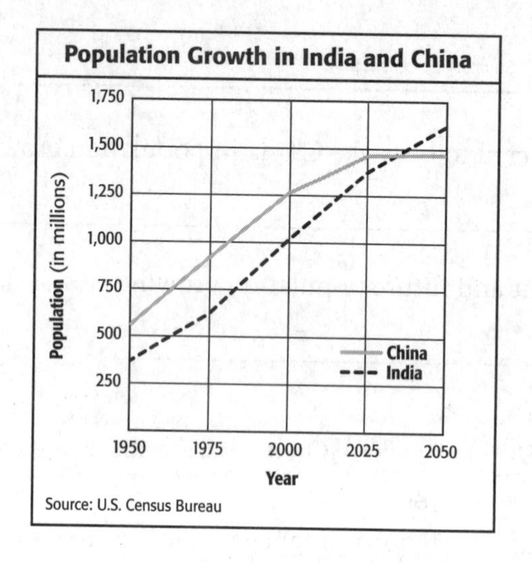

1. What is China's population expected to be in 2050?

2. Which country has a smaller population today?

COMPOSING AN ESSAY *(30 points)* On a separate sheet of paper, write an essay in response to *one* of the following.

1. Why have India's rivers and soils been called its most valuable resource? Give specific information to support your answer.

2. How have Bangladesh's geographic location and climate affected its development?

3. Discuss Britain's decision to partition India into the countries of India and Pakistan and the resulting problems that subdivided Pakistan.

4. Briefly discuss the major events and empires of India's early history to 1700.

Name ______________________ Class ______________ Date ______________

Chapter Test Form A

China, Mongolia, and Taiwan

REVIEWING FACTS

MATCHING *(3 points each)* In the space provided, write the letter of the term or place that matches each description. Some answers will not be used.

_____ **1.** The lowest point in China

_____ **2.** The harvesting of two crops a year from the same plot of land

_____ **3.** Capital city of Tibet

_____ **4.** A Chinese city taken by the British in 1842

_____ **5.** Simple pictures that represent ideas

_____ **6.** Taiwan's capital and largest city

_____ **7.** The first European colony in China

_____ **8.** Raising and harvesting fish in ponds or other bodies of water

_____ **9.** The Chang River runs through this region

_____ **10.** Large organized collective farms

a. Red Basin
b. double cropping
c. communes
d. martial law
e. Taipei
f. aquaculture
g. Hong Kong
h. Lhasa
i. Turpan Depression
j. pictograms
k. Macao
l. Yangtze

FILL IN THE BLANK *(3 points each)* For each of the following statements, fill in the blank with the appropriate word, phrase, or place.

1. The first dynasty in China was the ______________________.

2. The Great Leap Forward was a program to speed up ______________________ in China.

3. The Dalai Lama is the spiritual leader of ______________________.

4. Most Tibetans practice a form of ______________________ called Lamaism.

5. ______________________ is the official language of China.

6. ______________________ China is the country's most productive region.

7. The Plateau of Tibet has a dry ______________________ climate.

8. The ______________________ were the first Europeans to trade with China.

9. ______________________ is the least densely populated country in the world.

10. The last Chinese dynasty was overthrown by rebels under the leadership of

______________________.

UNDERSTANDING IDEAS *(3 points each)* For each of the following, write the letter of the best choice in the space provided.

_____ **1.** Which mountain range is not located in the northern part of China?
a. Greater Khingan Range
b. Tian Shan
c. Himalayas
d. Kunlun Shan

_____ **2.** Who was Deng Xiaoping?
a. Mao Zedong's successor
b. the head of the Chinese Nationalist Party
c. the head of the Taiwanese government for 38 years
d. Chiang Kai-shek's successor

_____ **3.** Which of the following is not an achievement of the Shang dynasty?
a. porcelain
b. chopsticks
c. shell money
d. musical instruments

_____ **4.** Which of the following is not a major environmental issue facing China today?
a. air pollution
b. desertification
c. flood control
d. soil erosion

_____ **5.** Where did the Japanese government set up a puppet government?
a. Tibet
b. Manchuria
c. Taiwan
d. Mongolia

Name ______________________ Class ______________ Date ______________

PRACTICING SKILLS *(5 points each)* Study the graph and answer the questions that follow.

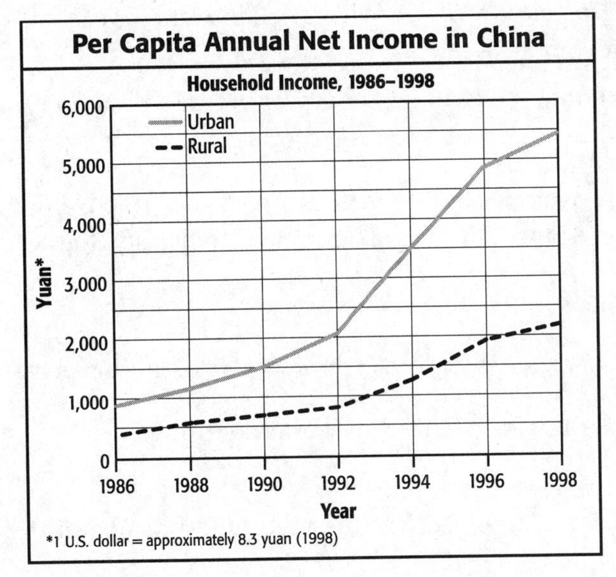

1. In which four-year period did the urban per capital annual net income increase the most?

____

2. Based on this graph, would you expect the gap between the urban and rural growth to increase or decrease in the future?

____

COMPOSING AN ESSAY *(15 points)* On a separate sheet of paper, write an essay in response to *one* of the following.

1. What are some benefits and drawbacks of China's Three Gorges Dam?

2. Where and when did Taoism originate? What do the followers of Taoism believe?

Name ______________________ Class ______________ Date ______________

Chapter Test Form B

China, Mongolia, and Taiwan

SHORT ANSWER *(10 points each)* Provide brief answers for each of the following. Use examples to support your answers.

1. Considering that only about 10 percent of China's land is arable, how has China managed to become the world's leading producer of many farm products?

2. Put the following dynasties in the order in which they ruled China, from earliest to latest: Han, Sung, Shang, Qin, T'ang. From which dynasty does China get its name?

3. Name the major rivers of China.

4. What are some characteristics of the Chinese style of landscape painting?

5. How is Chinese writing different from Western writing?

PRACTICING SKILLS *(10 points each)* Study the table and answer the questions that follow.

Productivity of China and Taiwan: per capita Gross Domestic Product

Year	China	Taiwan
1978	$440	$1,450
1983	$308	$2,673
1988	$320	$4,325
1993	$2,200	$10,600
1998	$3,600	$16,500

1. Which country's per capita GDP grew the most between 1978 and 1998? By about how many dollars did it grow?

2. In what five-year period did Taiwan experience the greatest increase in per capita GDP?

COMPOSING AN ESSAY *(30 points)* On a separate sheet of paper, write an essay in response to *one* of the following.

1. What was the Cultural Revolution?

2. How did Deng Xiaoping change China?

3. How has the economy of China changed since the death of Mao?

4. Explain what makes Taiwan one of Asia's richest and most industrialized countries.

Name ______________________ Class ______________ Date ____________

Chapter Test Form A

Japan and the Koreas

REVIEWING FACTS

MATCHING *(3 points each)* In the space provided, write the letter of the term or place that matches each description. Some answers will not be used.

_____ **1.** Highest peak in the Japanese Alps

_____ **2.** The elected law-making body of Japan

_____ **3.** Belief that work has moral value

_____ **4.** Largest and most populous of the Japanese islands

_____ **5.** Powerful warlord

_____ **6.** Financial support from the government

_____ **7.** Migration route for birds

_____ **8.** A truce

_____ **9.** Largest Ryukyu Island

_____ **10.** Densely populated region surrounding a central city

a. export economy
b. armistice
c. Honshū
d. Fuji
e. subsidies
f. work ethic
g. samurai
h. urban agglomeration
i. Okinawa
j. flyway
k. Diet
l. shogun

FILL IN THE BLANK *(3 points each)* For each of the following statements, fill in the blank with the appropriate word, phrase, or place.

1. ______________________ Korea has an export economy.

2. The ______________________ causes cool summers in the northern islands of Japan.

3. Japan controlled the southern half of ______________________ after the Russo-Japanese War.

4. ______________________ has one of the world's largest fishing fleets.

5. ______________________ is the leader of North Korea.

6. Japan's capital was moved from ______________________ to Tokyo during the Meiji Restoration.

7. Japan lies along a ______________________ zone.

8. Written Korean uses a 24-letter alphabet called ______________________.

9. Japan has a large trade ______________________ with the United States.

10. The early inhabitants of Japan were the ______________________.

UNDERSTANDING IDEAS *(3 points each)* For each of the following, write the letter of the best choice in the space provided.

______ **1.** What is a chaebol?
a. the Korean alphabet
b. a form of South Korean theater
c. a dish made with Chinese cabbage
d. a large family-owned conglomerate in South Korea

______ **2.** Japan and Russia disagree over ownership of which islands?
a. Sakhalin Islands
b. Ryukyu Islands
c. Okinawa
d. Kuril Islands

______ **3.** What climate does southern Japan have?
a. highland
b. humid subtropical
c. marine west coast
d. humid continental

______ **4.** Which of the following was not at some point occupied by Japan?
a. Taiwan
b. Mongolia
c. Korea
d. Manchuria

______ **5.** Which body of water does not border the Korean peninsula?
a. East China Sea
b. Yellow Sea
c. Sea of Japan
d. Korea Strait

PRACTICING SKILLS *(5 points each)* Study the map and answer the questions that follow.

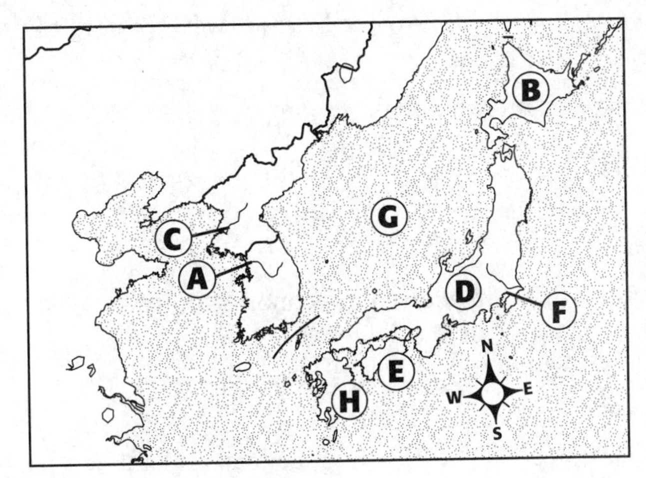

1. What island is marked by the letter B?

____

2. What city is marked by the letter C?

____

COMPOSING AN ESSAY *(15 points)* On a separate sheet of paper, write an essay in response to *one* of the following.

1. Compare and contrast tectonic activity in Japan and the Koreas.

2. Compare and contrast the economies of North and South Korea.

Name ______________________ Class ______________ Date ______________

Chapter Test Form B

Japan and the Koreas

SHORT ANSWER *(10 points each)* Provide brief answers for each of the following. Use examples to support your answers.

1. Name the four main islands that make up Japan.

2. Discuss how ocean currents affect Japan's climates. Be specific.

3. What is early Korean culture known for?

4. What characteristics of Japan allowed its economy to grow quickly?

5. How was Japan defeated in World War II?

PRACTICING SKILLS *(10 points each)* Study the map and answer the questions that follow.

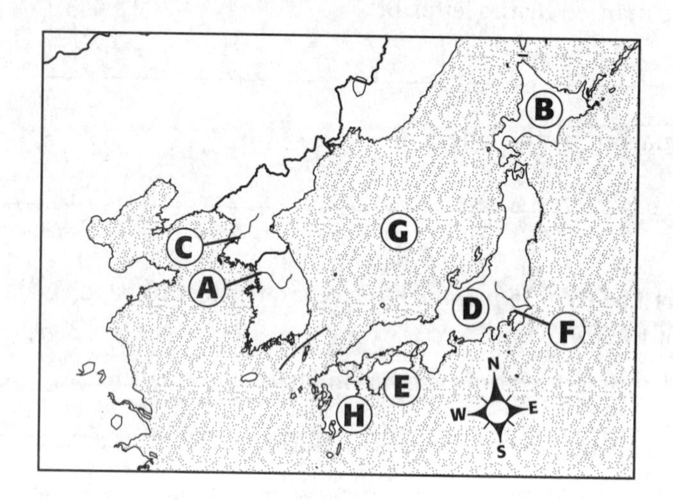

1. What body of water is marked by the letter G?

2. What island is marked by the letter H?

COMPOSING AN ESSAY *(30 points)* On a separate sheet of paper, write an essay in response to *one* of the following.

1. Summarize the history of Korea since 1945.

2. How are the cultures of Japan and Korea similar?

3. Describe the political system of Japan in the 1100s.

4. What have been the benefits and disadvantages of the keiretsu system for Japan's economy?

Name ______________________ Class ______________ Date ______________

Chapter Test Form A

Mainland Southeast Asia

REVIEWING FACTS

MATCHING *(3 points each)* In the space provided, write the letter of the term or place that matches each description. Some answers will not be used.

______ **1.** Largest city and capital of Myanmar

______ **2.** A temple complex built by the Khmer

______ **3.** A Buddhist temple and monastery

______ **4.** A river that flows across northern Thailand into the Gulf of Tonkin

______ **5.** A narrow peninsula that includes part of Thailand

______ **6.** Southeast Asia's largest freshwater lake

______ **7.** Capital city of Thailand

______ **8.** The idea that if one country converts to communism then others will follow

______ **9.** Network of canals in Bangkok

______ **10.** Tree-dwelling

a. Indochina Peninsula
b. domino theory
c. Yangon
d. Malay Peninsula
e. klongs
f. Hong River
g. Tonle Sap
h. Bangkok
i. wat
j. arboreal
k. Irrawaddy River
l. Angkor Wat

FILL IN THE BLANK *(3 points each)* For each of the following statements, fill in the blank with the appropriate word, phrase, or place.

1. ______________________ is the most industrialized and economically advanced country in the region.

2. The driest area in the region is the ______________________ in Thailand.

3. The former name of Thailand is ______________________.

4. ______________________ is still recovering from the repressive rule of the Khmer Rouge.

5. Subsistence farmers have been discouraged from practicing ______________________ agriculture because of the potential damage to the rain forest.

6. ______________________ was the first religion introduced into the region.

7. The capital of Vietnam is ______________________.

8. The most popular type of Buddhism in the region is ______________________.

9. ______________________ is the only landlocked country in mainland Southeast Asia.

10. ______________________ is a type of soil commonly found in rain forests that becomes useless when the rain forest cover is removed.

UNDERSTANDING IDEAS *(3 points each)* For each of the following, write the letter of the best choice in the space provided.

______ **1.** Thailand is the leading producer of
- **a.** rubber.
- **b.** opium.
- **c.** copper.
- **d.** rice.

______ **2.** Which country has been most influenced by China?
- **a.** Laos
- **b.** Vietnam
- **c.** Cambodia
- **d.** Myanmar

______ **3.** Which of the following is not a way that rivers have influenced the region?
- **a.** provided fertile alluvial soil
- **b.** proved useful for local transportation
- **c.** influenced the growth and location of cities
- **d.** determined national boundaries

______ **4.** What types of climates are found in mainland Southeast Asia?
- **a.** tropical and subtropical
- **b.** tropical and highland
- **c.** subtropical and Mediterranean
- **d.** tropical and marine west coast

______ **5.** What country invaded Cambodia in 1979, thus ending the Khmer Rouge government?
- **a.** Myanmar
- **b.** Laos
- **c.** Vietnam
- **d.** Thailand

Name ______________________ Class ______________ Date ____________

PRACTICING SKILLS *(5 points each)* Study the pie charts and answer the questions that follow.

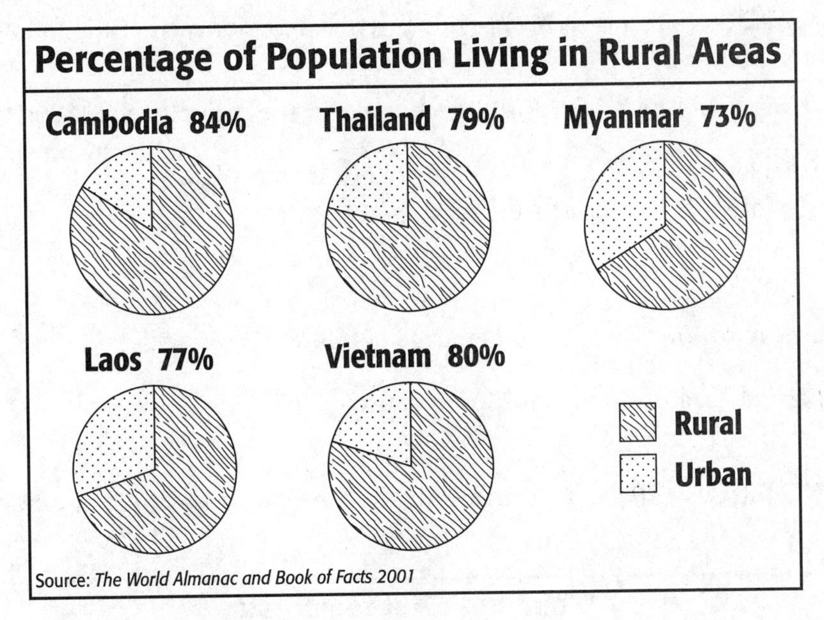

1. Which country is the most highly urbanized?

____

2. Which two countries have the most similar ratio of urban to rural populations?

____

COMPOSING AN ESSAY *(15 points)* On a separate sheet of paper, write an essay in response to *one* of the following.

1. Briefly describe the influence and history of France in mainland Southeast Asia.

2. Discuss the problems that have been caused by the clearing of tropical rain forests in the region.

Name ______________________ Class ____________ Date ____________

Chapter Test Form B

Mainland Southeast Asia

SHORT ANSWER *(10 points each)* Provide brief answers for each of the following. Use examples to support your answers.

1. What changes did the British and French make in the region in the 1800s?

2. What are mainland Southeast Asia's four major rivers? What soils in the river valleys and deltas support intensive farming?

3. What is unusual about Tonle Sap?

4. What are the region's three main language families?

5. What natural hazards do the monsoons cause?

PRACTICING SKILLS *(10 points each)* Study the table and answer the questions that follow.

Mainland Southeast Asia

Country	Population	Growth Rate	Literacy Rate	Per Capita GDP
Cambodia	11,626,520	2.5%	65%	$700
Laos	5,407,453	2.7%	60%	$1,260
Myanmar	48,081,302	1.6%	83%	$1,200
Thailand	60,609,046	0.9%	94%	$6,100
Vietnam	77,311,210	1.4%	94%	$1,770

1. Which country has the highest growth rate but the smallest population?

2. Which country has a per capita GDP that is more than three times higher than any of the other countries?

COMPOSING AN ESSAY *(30 points)* On a separate sheet of paper, write an essay in response to *one* of the following.

1. How does Buddhism shape the lives of the region's people?

2. What is the history of the Khmer in Cambodia?

3. Why did the United States become involved in the Vietnam War?

4. How do the cultures of urban and rural areas in mainland Southeast Asia differ?

Name ______________________ Class ____________ Date ____________

Chapter Test Form A

Island Southeast Asia

REVIEWING FACTS

MATCHING *(3 points each)* In the space provided, write the letter that matches each description. Choose your answers from the list below. Some answers will not be used.

_____ **1.** Of the same kind

_____ **2.** Volcanic mudflows

_____ **3.** Cooperative economic development group

_____ **4.** A form of shifting cultivation

_____ **5.** Animals or plants that are native to a certain area

_____ **6.** A city state in island Southeast Asia

_____ **7.** Capital city of the Philippines

_____ **8.** A large group of islands

_____ **9.** Largest, earliest kingdom in island Southeast Asia

_____ **10.** Villages built on stilts

a. slash-and-burn
b. kampongs
c. endemic species
d. Manila
e. archipelago
f. homogeneous
g. Singapore
h. Majapahit
i. Jakarta
j. lahars
k. Kuala Lumpur
l. ASEAN

FILL IN THE BLANK *(3 points each)* For each of the following statements, fill in the blank with the appropriate word, phrase, or place.

1. The major industry on the island of Bali is ______________________.

2. Culturally and ethnically, the most homogeneous country in island Southeast Asia is ______________________.

3. ______________________ is a fabric on which wax is applied to the cloth to create designs.

4. With the exception of ______________________, most countries in island Southeast Asia are poor.

5. The major movement of people in island Southeast Asia is toward ______________________.

6. The most productive method of rice cultivation is ______________________.

7. Manila, Philippines, and ______________________, Indonesia, are capital cities with large slums.

8. ______________________ is the world's third largest island.

9. The Philippines were first colonized by ______________________.

10. ______________________ is a tiny country ruled by a sultan.

UNDERSTANDING IDEAS *(3 points each)* For each of the following, write the letter of the best choice in the space provided.

_____ **1.** The Strait of Malacca lies between Sumatra and
a. the Malay Peninsula.
b. the Philippines.
c. Indonesia.
d. Bali.

_____ **2.** Which of the following is not an official language of Singapore?
a. Tagalog
b. Chinese
c. Malay
d. English

_____ **3.** Religions commonly practiced in island Southeast Asia include all of the following except
a. Buddhism.
b. Hinduism.
c. Confucianism.
d. Islam.

_____ **4.** Malaysia and Indonesia are among the world's largest producers of
a. spices.
b. natural rubber.
c. oil and natural gas.
d. coffee.

_____ **5.** Where were United Nations peace-keeping troops deployed in 1999?
a. East Timor
b. Malaysia
c. Singapore
d. Philippines

Name ______________________ Class ______________ Date ______________

PRACTICING SKILLS *(5 points each)* Study the diagram and answer the questions that follow.

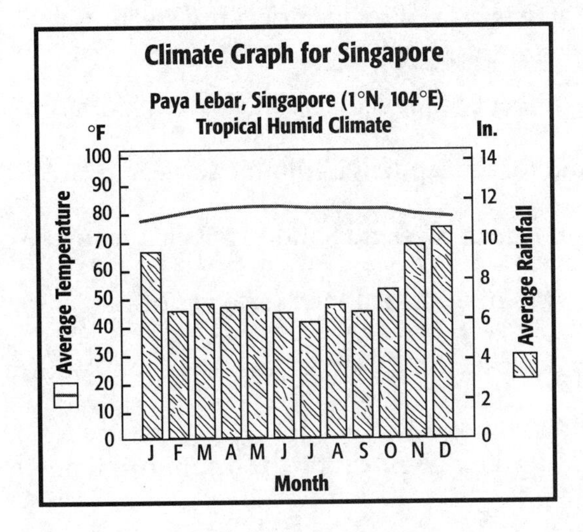

1. What is the average annual rainfall for Singapore?

____

____

2. Compare and contrast the annual variation in temperature and rainfall.

____

____

COMPOSING AN ESSAY *(15 points)* On a separate sheet of paper, write an essay in response to *one* of the following.

1. Explain how island Southeast Asia's location, resources, and climate make it a promising area for economic growth.

2. How has the diversity of religious and ethnic groups in island Southeast Asia affected the area?

Name ______________________ Class ______________ Date ____________

Chapter Test Form B

Island Southeast Asia

SHORT ANSWER *(10 points each)* Provide brief answers for each of the following. Use examples to support your answers.

1. What three factors influence the region's climates and biomes?
2. What factors account for Singapore's economic success?
3. Why were Europeans drawn to island Southeast Asia during the colonial era?
4. What physical processes have shaped the landforms of island Southeast Asia?
5. Identify and briefly describe three ways farmers in island Southeast Asia grow rice.

PRACTICING SKILLS *(10 points each)* Study the diagram and answer the questions that follow.

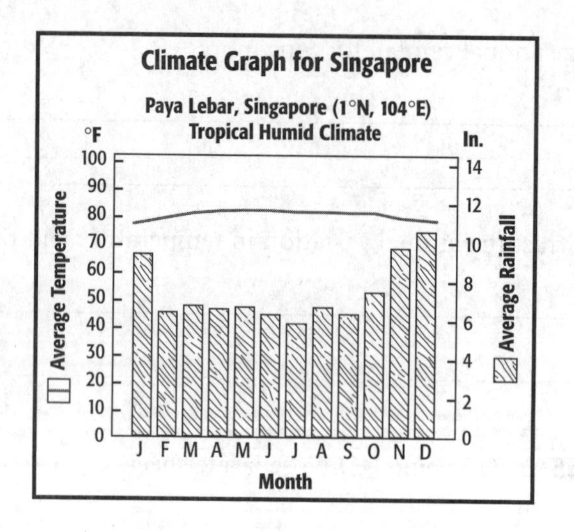

1. What month receives the least amount of rain?
2. What evidence on this graph suggests that Singapore has a tropical humid climate?

COMPOSING AN ESSAY *(30 points)* On a separate sheet of paper, write an essay in response to *one* of the following.

1. Why are some countries in island Southeast Asia known as the Tigers of the Pacific Rim?
2. Compare and contrast Singapore and Brunei.
3. How are human activities endangering the region's forests and animals?
4. Explain why the people and culture of the Philippines are unique in the region.

Name ______________________ Class ______________ Date ____________

Unit Test Form A

East and Southeast Asia

REVIEWING FACTS

MATCHING *(3 points each)* In the space provided, write the letter of the term or place that matches each description. Some answers will not be used.

_____ **1.** A large group of islands
_____ **2.** Powerful warlord
_____ **3.** A temple complex built by the Khmer
_____ **4.** The first European colony in China
_____ **5.** Volcanic mudflows
_____ **6.** Tree-dwelling
_____ **7.** Largest and most populous of the Japanese islands
_____ **8.** Simple pictures that represent ideas
_____ **9.** Animals or plants that are native to a certain area
_____ **10.** The lowest point in China

a. Angkor Wat
b. Macao
c. Honshū
d. Tonle Sap
e. Turpan Depression
f. lahars
g. shogun
h. archipelago
i. pictograms
j. domino theory
k. arboreal
l. endemic species

FILL IN THE BLANK *(3 points each)* For each of the following statements, fill in the blank with the appropriate word, phrase, or place.

1. ______________________ is the leader of North Korea.

2. ______________________ is still recovering from the repressive rule of the Khmer Rouge.

3. The first dynasty in China was the ______________________.

4. The earliest inhabitants of Japan were the ______________________.

5. Culturally and ethnically, the most homogeneous country in island Southeast Asia is ______________________.

6. The Plateau of Tibet has a dry ______________________ climate.

7. With the exception of ______________________, most countries in island Southeast Asia are poor.

8. The Dalai Lama is the spiritual leader of ______________________.

9. The Philippines were first colonized by ____________________.

10. ______________________ is the only landlocked country in mainland Southeast Asia.

UNDERSTANDING IDEAS *(3 points each)* For each of the following, write the letter of the best choice in the space provided.

_____ **1.** What types of climates are found in mainland Southeast Asia?
a. tropical and subtropical
b. tropical and highland
c. subtropical and Mediterranean
d. tropical and marine west coast

_____ **2.** Which body of water does not border the Korean peninsula?
a. East China Sea
b. Yellow Sea
c. Sea of Japan
d. Korea Strait

_____ **3.** What is a chaebol?
a. the Korean alphabet
b. a form of South Korean theater
c. a dish made with Chinese cabbage
d. a large family-owned conglomerate in South Korea

_____ **4.** Malaysia and Indonesia are among the world's largest producers of
a. spices.
b. natural rubber.
c. oil and natural gas.
d. coffee.

_____ **5.** Where did the Japanese government set up a puppet government?
a. Tibet
b. Manchuria
c. Taiwan
d. Mongolia

Name ______________________ Class ______________ Date ______________

PRACTICING SKILLS *(5 points each)* Study the diagram and answer the questions that follow.

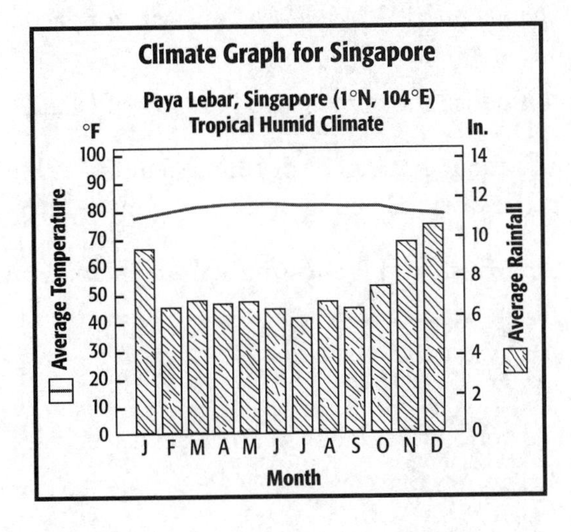

1. What is the annual rainfall for Singapore?

____

2. Compare and contrast the annual variation in temperature and rainfall.

____

COMPOSING AN ESSAY *(15 points)* On a separate sheet of paper, write an essay in response to one of the following.

1. How has the diversity of religious groups in island Southeast Asia affected the area?

2. Compare and contrast the economies of North and South Korea.

Name ______________________ Class ______________ Date ______________

Unit Test Form B

East and Southeast Asia

SHORT ANSWER *(10 points each)* Provide brief answers for each of the following. Use examples to support your answers.

1. What three factors influence the climates and biomes of island Southeast Asia?

2. Considering that only about 10 percent of China's land is arable, how has China managed to become the world's leading producer of many farm products?

3. What changes did the British and French make in mainland Southeast Asia in the 1800s?

4. Identify and briefly describe three ways farmers in island Southeast Asia grow rice.

5. Discuss how ocean currents affect Japan's climates. Be specific.

PRACTICING SKILLS *(10 points each)* Study the diagram and answer the questions that follow.

Productivity of China and Taiwan: per capita Gross Domestic Product

Year	China	Taiwan
1978	$440	$1,450
1983	$308	$2,673
1988	$320	$4,325
1993	$2,200	$10,600
1998	$3,600	$16,500

1. Which country's per capita GDP grew the most between 1978 and 1998? By about how many dollars did it grow?

2. In what five-year period did Taiwan experience the greatest rate of growth in per capita GDP?

COMPOSING AN ESSAY *(30 points)* On a separate sheet of paper, write an essay in response to *one* of the following.

1. Compare and contrast the cultures of Japan and Korea.

2. How does Buddhism shape the lives of the people of mainland Southeast Asia?

3. Why are some countries in island Southeast Asia known as Tigers of the Pacific Rim?

4. What was the Cultural Revolution?

Name ______________________ Class ______________ Date ____________

Chapter Test Form A

Australia and New Zealand

REVIEWING FACTS

MATCHING *(3 points each)* In the space provided, write the letter that matches each description. Choose your answers from the list below. Some answers will not be used.

_____ **1.** Requires the use of a large amount of land but little capital and labor per unit area

_____ **2.** Capital of New Zealand

_____ **3.** Plants and animals that are not native to an area

_____ **4.** Ability to produce goods in sufficient numbers to reduce unit cost

_____ **5.** Wells in which water flows naturally to the surface

_____ **6.** The first people to settle New Zealand

_____ **7.** Mammals that have pouches to carry their young

_____ **8.** A large, flightless bird that is now extinct

_____ **9.** Capital of Australia

_____ **10.** Dry interior region of Australia

a. marsupials
b. Canberra
c. exotic species
d. Aborigines
e. Wellington
f. extensive agriculture
g. Auckland
h. economy of scale
i. outback
j. Maori
k. Moa
l. artesian wells

FILL IN THE BLANK *(3 points each)* For each of the following statements, fill in the blank with the appropriate word, phrase, or place.

1. Ranchers in Australia and New Zealand raise mainly ______________________ and ______________________.

2. The only mammals native to New Zealand are ______________________.

3. Australia's main mountain system is the ______________________.

4. Most of the population of New Zealand lives on ______________________ Island.

5. Aborigines believe that their ancestors created the world during the ______________________.

6. The North and South Islands of New Zealand are separated by the ______________________.

7. About 75 percent of Australia's ______________________ have been destroyed in the last 20 years.

8. The Maori Wars of 1845–1872 were won by ______________________.

9. Australia exports mainly ______________________ materials.

10. New Zealand contains active volcanoes because it is located on the

______________________.

UNDERSTANDING IDEAS *(3 points each)* For each of the following, write the letter of the best choice in the space provided.

______ **1.** Which of the following is not a factor in the Australian climate?
- **a.** the Great Dividing Range
- **b.** the Great Barrier Reef
- **c.** its latitude
- **d.** low elevation

______ **2.** What is the main source of electricity for New Zealand?
- **a.** nuclear
- **b.** wind
- **c.** geothermal
- **d.** hydroelectric

______ **3.** The major industries of New Zealand include all of the following except
- **a.** manufacturing.
- **b.** tourism.
- **c.** farming.
- **d.** mining.

______ **4.** What is the climate of New Zealand?
- **a.** tropical wet and dry
- **b.** humid continental
- **c.** marine west coast
- **d.** humid subtropical

______ **5.** When were the first British settlements established in Australia?
- **a.** 1830
- **b.** 1788
- **c.** 1851
- **d.** 1901

PRACTICING SKILLS *(5 points each)* Study the maps and answer the questions that follow.

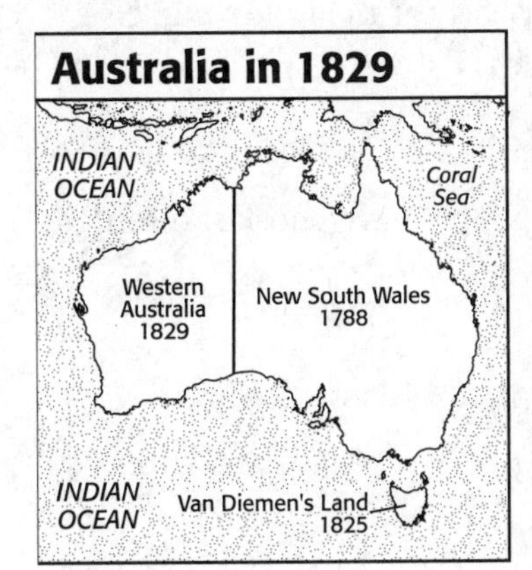

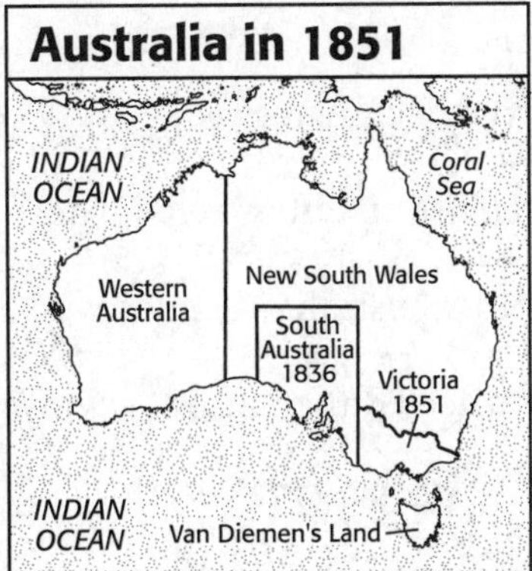

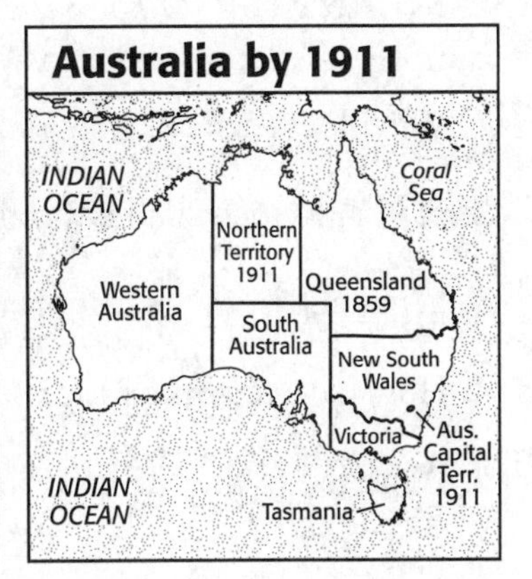

1. Using these maps and your knowledge of the geography of Australia, explain the population growth of Australia.

____

____

2. What do the names of the Australian states suggest about its history as a colony?

____

____

COMPOSING AN ESSAY *(15 points)* On a separate sheet of paper, write an essay in response to *one* of the following.

1. Compare and contrast the colonization of New Zealand and Australia.

2. How have recent changes in Europe affected the economies of New Zealand and Australia?

Name ______________________ Class ______________ Date ____________

Chapter Test Form B

Australia and New Zealand

SHORT ANSWER *(10 points each)* Provide brief answers for each of the following. Use examples to support your answers.

1. Why does Australia have such dry climates?

2. How has Australia's natural environment affected where people have chosen to settle?

3. What activities are important to Australia's economy?

4. How have tectonic processes affected the natural environment of New Zealand?

5. What led to the Maori Wars of 1845–1872?

PRACTICING SKILLS *(10 points each)* Study the maps and answer the questions that follow.

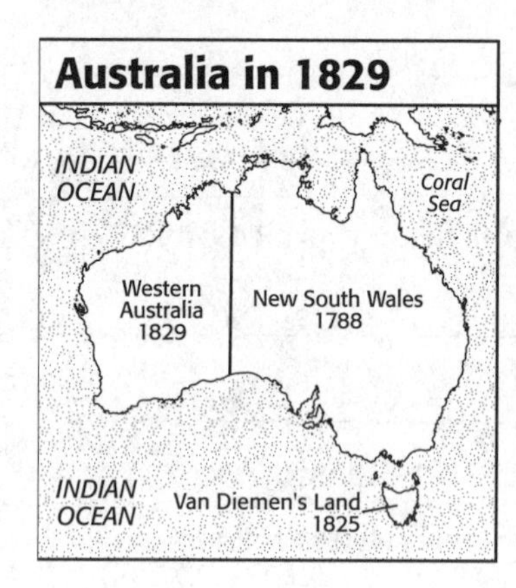

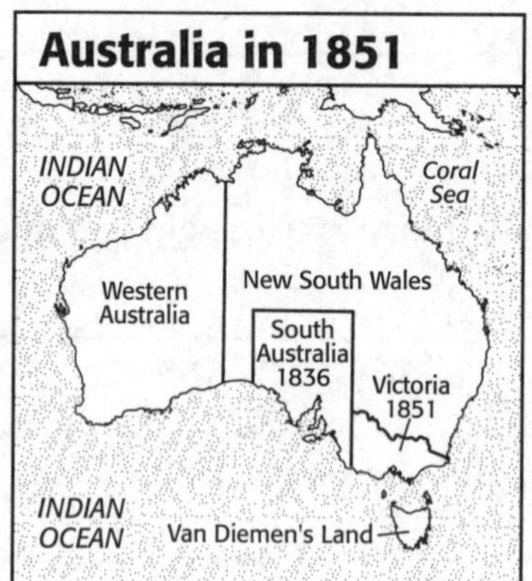

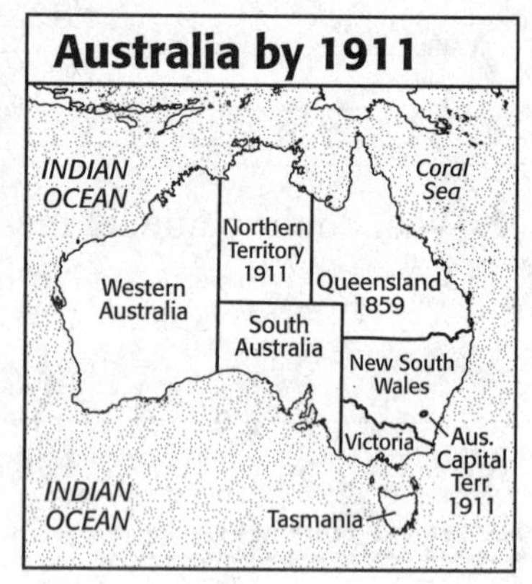

1. Based on the settlement shown on these maps, what would you expect the physical terrain and climate of Western Australia to be like?

2. How many states are found in what was once New South Wales?

COMPOSING AN ESSAY *(30 points)* On a separate sheet of paper, write an essay in response to *one* of the following.

1. How are New Zealand's major commercial industries tied to its agricultural products?

2. Describe both the strengths and weaknesses of the Australian economy. How is the Australian government addressing these weaknesses?

3. How did the British colonization of Australia and New Zealand differ?

4. Compare and contrast the North and South Islands of New Zealand.

Name ______________________ Class ______________ Date ____________

Chapter Test Form A

The Pacific Islands

REVIEWING FACTS

MATCHING *(3 points each)* In the space provided, write the letter of the term or place that matches each description. Some answers will not be used.

______ **1.** Societies that trace kinship through the mother

______ **2.** Dried coconut meat

______ **3.** The capital of Fiji

______ **4.** Areas placed under the control of another territory while they prepare for independence

______ **5.** An island that is still a U.S. territory

______ **6.** Subregion of the Pacific Islands closest to Australia

______ **7.** Chemicals used to make fertilizers

______ **8.** A ring-shaped coral island

______ **9.** Area where the prevailing winds meet near the equator

______ **10.** Subregion of the Pacific Islands east of the Philippines

a. Intertropical Convergence Zone
b. matrilineal
c. trust territories
d. Melanesia
e. Port Moresby
f. atoll
g. copra
h. Suva
i. Fiji
j. phosphates
k. Wake Island
l. Micronesia

FILL IN THE BLANK *(3 points each)* For each of the following statements, fill in the blank with the appropriate word, phrase, or place.

1. ______________________ islands form from coral and are usually small and flat.

2. The main source of protein in the region is ______________________.

3. ______________________ islands are formed from volcanic mountains that have grown from the ocean floor to the surface.

4. The populations of most Pacific Islands have a very ______________________ rate of natural increase.

5. ______________________ is the most culturally distinct of the three subregions in the Pacific Islands.

6. ______________________ was the first fuel Europeans sought in the region.

7. Of the Pacific islands, only ______________________ has a highland climate.

8. Wake Island, American Samoa, and ______________________ are still U.S. territories.

9. ______________________ continued to use its Pacific territories as testing grounds for nuclear weapons well in to the 1990s.

10. The mining of phosphates is a major economic activity in

______________________.

UNDERSTANDING IDEAS *(3 points each)* For each of the following, write the letter of the best choice in the space provided.

______ **1.** Islands that lie on a shallow continental shelf are called
- **a.** atolls.
- **b.** oceanic islands.
- **c.** continental islands.
- **d.** volcanic islands.

______ **2.** Low islands are formed from
- **a.** copra.
- **b.** coral.
- **c.** phosphates.
- **d.** volcanic ash.

______ **3.** How far out from an island does its Exclusive Economic Zone extend?
- **a.** 100 miles
- **b.** 200 miles
- **c.** 300 miles
- **d.** 400 miles

______ **4.** Which of the following is not a major root crop of the region?
- **a.** sweet potatoes
- **b.** yams
- **c.** taro
- **d.** plantains

______ **5.** What causes ocean trenches?
- **a.** subduction
- **b.** earthquakes
- **c.** volcanoes
- **d.** erosion

PRACTICING SKILLS *(10 points each)* Study the map and answer the questions that follow.

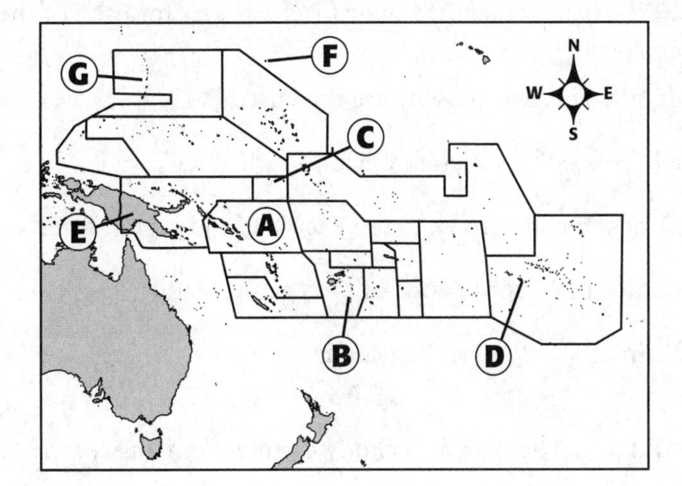

1. What letter represents Nauru?

____

2. What letter represents the Solomon Islands?

____

COMPOSING AN ESSAY *(15 points)* On a separate sheet of paper, write an essay in response to *one* of the following.

1. Discuss how migration, war, and trade have affected the people and cultures of the region.

2. What are some challenges facing the Pacific Islands?

Name ______________________ Class ______________ Date ______________

Chapter Test Form B

The Pacific Islands

SHORT ANSWER *(10 points each)* Provide brief answers for each of the following. Use examples to support your answers.

1. What are pidgin languages and why are they useful?

2. What are the benefits of an Exclusive Economic Zone?

3. What environmental concerns does the Pacific Island region face?

4. What are the three main subregions of the Pacific Islands?

5. Describe the climates of the Pacific Islands.

PRACTICING SKILLS *(10 points each)* Study the map and answer the questions that follow.

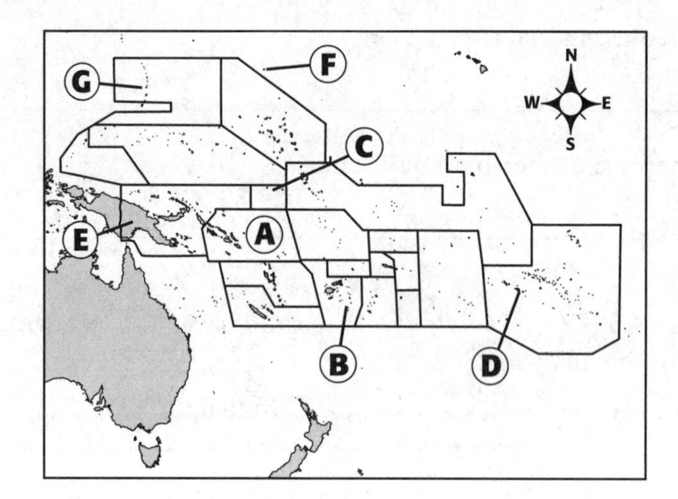

1. What letter represents Wake Island?

2. What letter represents Tahiti?

COMPOSING AN ESSAY *(30 points)* On a separate sheet of paper, write an essay in response to *one* of the following.

1. Discuss the two types of Pacific islands and how they are formed. How viable is each as a place for humans to live?

2. How are migration patterns and the general economic status of the region related?

3. Discuss the importance of coral reefs to the diversity of the region.

4. Describe how the Pacific Islands were first settled.

Name ______________________ Class ______________ Date ____________

Unit Test Form A

The Pacific World

REVIEWING FACTS

MATCHING *(3 points each)* In the space provided, write the letter of the term or place that matches each description. Some answers will not be used.

______ **1.** A ring-shaped coral island

______ **2.** The capital of New Zealand

______ **3.** Plants and animals that are not native to an area

______ **4.** Dried coconut meat

______ **5.** Well in which water flows naturally to the surface

______ **6.** Chemicals used to make fertilizers

______ **7.** Mammals that have pouches to carry their young

______ **8.** Areas placed under control of another country while they prepare for independence

______ **9.** The capital of Australia

______**10.** Societies that trace kinship through the mother

a. matrilineal
b. Canberra
c. exotic species
d. phosphates
e. Wellington
f. marsupials
g. trust territories
h. Port Moresby
i. Aborigines
j. atoll
k. copra
l. artesian well

FILL IN THE BLANK *(3 points each)* For each of the following statements, fill in the blank with the appropriate word, phrase, or place.

1. Large cattle ranches in Australia are an example of ______________________ farming.

2. The populations of most Pacific Islands have a very ______________________ rate of natural increase.

3. Australia's main mountain system is the ______________________.

4. ______________________ was the first fuel Europeans sought in the region.

5. Aborigines believe that their ancestors created the world during the ______________________.

6. ______________________ continued to use its Pacific territories as testing grounds for nuclear weapons well into the 1990s.

7. Northeast and southeast ______________________ meet at the Intertropical Convergence Zone.

8. Australia exports mainly ______________________ materials.

9. ______________________ is the most culturally distinct of the three subregions in the Pacific Islands.

10. New Zealand contains active volcanoes because it is located on the

______________________.

UNDERSTANDING IDEAS *(3 points each)* For each of the following, write the letter of the best choice in the space provided.

______ **1.** Who was the first European explorer to land in New Zealand?
a. Tasman
b. Cook
c. Magellan
d. Van Diemen

______ **2.** Islands that lie on a shallow continental shelf are called
a. atolls.
b. oceanic islands.
c. continental islands.
d. volcanic islands.

______ **3.** All of the following are major industries in New Zealand except
a. manufacturing.
b. tourism.
c. farming.
d. mining.

______ **4.** What causes ocean trenches?
a. subduction
b. earthquakes
c. volcanoes
d. erosion

______ **5.** Which of the following is not a landform of Australia?
a. Great Dividing Range
b. Southern Alps
c. Central Lowlands
d. Western Plateau

Name ______________________ Class ______________ Date ______________

PRACTICING SKILLS *(5 points each)* Study the maps and answer the questions that follow.

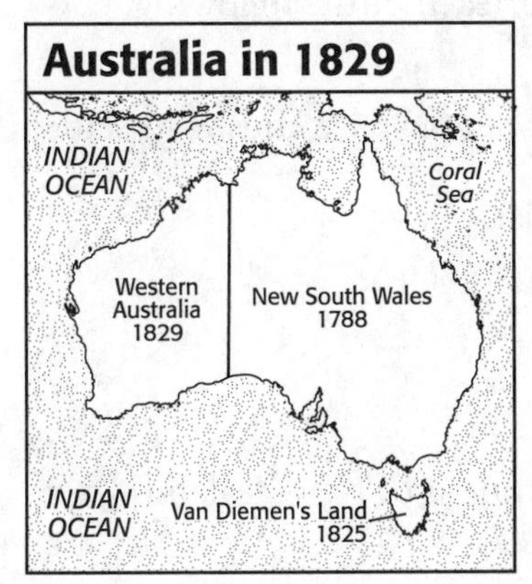

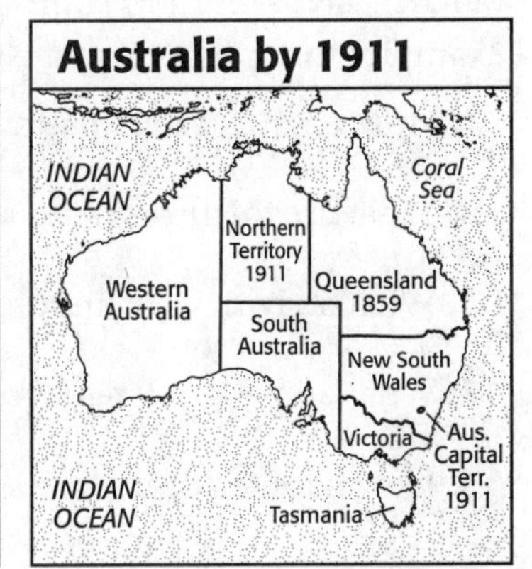

1. Using maps and your knowledge of the geography of Australia, explain the population growth of Australia.

____

2. What do the names of the Australian states reveal about its history as a colony?

____

COMPOSING AN ESSAY *(15 points)* On a separate sheet of paper, write an essay in response to *one* of the following.

1. Discuss how migration, war, and trade have affected the people and cultures of the Pacific Islands.

2. How have recent changes in Europe affected the economies of New Zealand and Australia?

Name ______________________ Class ______________ Date ____________

Unit Test Form B

The Pacific World

SHORT ANSWER *(10 points each)* Provide brief answers for each of the following. Use examples to support your answers.

1. How has Australia's natural environment affected where people have chosen to settle?
2. What are the three main subregions of the Pacific Islands?
3. What activities are important to Australia's economy?
4. What are pidgin languages and why are they useful?
5. How have tectonic processes affected the natural environment of New Zealand?

PRACTICING SKILLS *(10 points each)* Study the maps and answer the questions that follow.

Australia in 1829

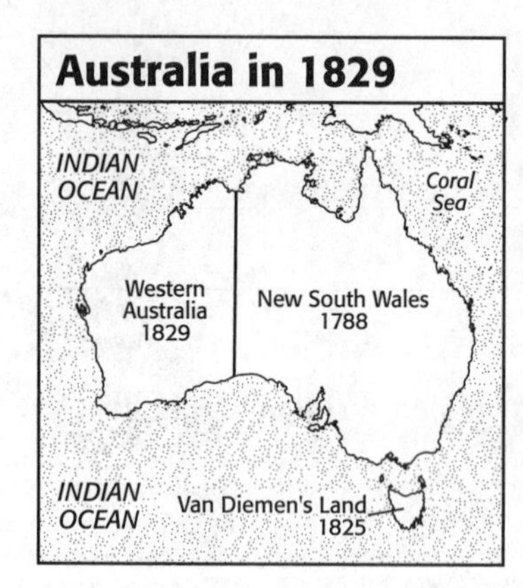

Australia in 1851

Australia by 1911

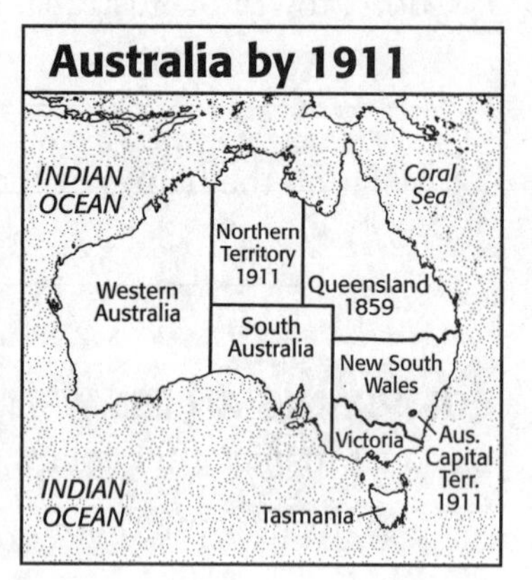

1. Based on the settlement shown on these maps, what would you expect the physical terrain and climate of Western Australia to be like?
2. How many states are now found in what was once New South Wales?

COMPOSING AN ESSAY *(30 points)* On a separate sheet of paper, write an essay in response to *one* of the following.

1. Compare and contrast the North and South Islands of New Zealand.
2. Discuss the importance of coral reefs to the diversity of the Pacific.
3. Compare and contrast the colonization of Australia and New Zealand.
4. Discuss the two types of Pacific islands and how they are formed. How viable is each as a place for humans to live?

Chapter 1

FORM A

Matching

1. c
2. f
3. e
4. g
5. a
6. b
7. d
8. l
9. k
10. i

Fill in the Blank

1. equator
2. human
3. legend or key
4. Indian
5. compass rose
6. movement
7. population
8. great-circle
9. contour
10. flat-plane

Understanding Ideas

1. c
2. c
3. b
4. b
5. b

Practicing Skills

1. 159° 58' E, 9° 25' S
2. 2,000 ft.

Composing an Essay

1. Answers may vary but should mention the three types of maps: cylindrical, conic, and flat plane. Answers should mention that cylindrical maps distort the sizes of landmasses at the poles but give true direction and shape of the continents; conic maps show the true shapes of landmasses along the lines of latitude that they touch; and flat-plane projections show the true sizes of landmasses but distort their shapes. Flat-plane maps are useful for navigators and pilots, the others are useful for reference and visual understanding of landmasses.
2. Answers may vary but should mention that geography helps people understand why things are located where they are and how people interact with their environments. Maps facilitate accurate and timely travel and are useful for making business projections and determining business locations. Maps and other geographic tools can demonstrate the affect of development on the environment.

FORM B

Short Answer

1. location, place, human-environment interaction, movement, and region
2. The three types of regions are functional, perceptual, and formal. Examples might include the Amtrak train system as a functional region, the Sun Belt as a perceptual region, and the Appalachian Mountains as a formal region.
3. Answers will vary but might mention land developers, military, and schools.
4. Answers will vary. Maps described may include precipitation, climate, population, elevation, and contour maps.
5. Geographers have superimposed a grid on the globe that includes lines of latitude and longitude. Lines of latitude run from east to west while lines of longitude run from north to south. These lines are measured in degrees that are further divided into minutes. The entire circle of the globe equals 360 degrees. Geographers have also organized the world into hemispheres, oceans, and continents.

Practicing Skills

1. Guadalcanal appears to be an extremely mountainous island with a flat plain near sea level located on the north side of the island.
2. The sharpest drop in elevation occurs between Mt. Makarakomburu and the northern beach area, where the elevation drops from greater than 8,000 feet to sea level in a relatively short space.

Composing an Essay

1. Answers will vary but should mention that human geography looks at the distribution and characteristics of the world's people, including where they live and work as well as their ways of life. Physical geography focuses on Earth's natural environments and the processes that affect them.
2. Answers will vary but should mention the six essential elements: the world in spatial terms, places and regions, physical systems, human systems, environment and society, and the uses of geography. These essential elements share many properties

with the five themes but provide a way of linking different themes together to make a more seamless study of geography.

3. Answers will vary but should mention that the three types of map projections—conic, cylindrical, and flat-plane—serve different purposes. It is not possible to create a two-dimensional map that can accurately portray shape, size, and distance of a three-dimensional object such as Earth. Each type of map seeks to depict at least one aspect in an accurate fashion.

4. Answers will vary but should mention that cartographers rely heavily on computers to actually draw maps today, whereas in the past maps were drawn by hand. The ability to input accurate measurements in a computer and have it render an accurate representation is a huge advancement. The use of satellite photographs allows cartographers to obtain accurate measurements.

Chapter 2

FORM A

Matching

1. h
2. e
3. j
4. b
5. a
6. k
7. i
8. l
9. g
10. c

Fill in the Blank

1. reflect
2. spring
3. water
4. 100
5. elliptical
6. lithosphere
7. North Star
8. gravity
9. $365\frac{1}{4}$
10. Tides

Understanding Ideas

1. c
2. d
3. c
4. d
5. c

Practicing Skills

1. Earth's tilt affects the amount of solar energy received by various places at various times of the year, thereby creating seasons.

2. June 21

Composing an Essay

1. Answers may vary but should mention that having a complete understanding of the interactions of the biosphere, lithosphere, atmosphere, and hydrosphere allows people to make informed decisions regarding the environmental impact of human or physical changes.

2. Answers may vary but should note that day would last six months and night would last six months, thus radically changing the climate of the world.

FORM B

Short Answer

1. The atmosphere is the envelope of gases that surrounds Earth. The lithosphere is the solid crust of the planet. The hydrosphere includes all of Earth's water. The biosphere is the part of Earth that includes all life forms.

2. The major bodies that make up the solar system are the Sun, the nine planets, asteroids, comets, and moons.

3. During the December solstice, the South Pole is tilted toward the Sun. During the June solstice, the North Pole is pointed toward the Sun.

4. The Moon has a barren volcanic surface and many craters that were created by the impact of meteors and comets. The Moon has no air, water, or life.

5. Low-latitude areas are located near the equator, which receives a lot of solar energy all year. Polar regions receive very little solar energy.

Practicing Skills

1. The tilt of Earth's axis in relation to the Sun causes the seasons.

2. summer

Composing an Essay

1. Answers may vary but should mention and define rotation, revolution, and tilt. Answers should note the degree of tilt in the Earth's axis as well as the elliptical orbit of Earth over a period of $365\frac{1}{4}$ days.

2. Answers may vary but should mention that solstices and equinoxes share some characteristics—both occur twice a year, six months apart, while one marks summer and winter and the other spring and fall. However, solstices mark the time that Earth's poles are at their greatest angle toward or away from the Sun, while equinoxes mark the time when Earth's poles are perpendicular to the Sun.

3. Answers may vary but should mention that the inner planets are closest to the Sun and are terrestrial, meaning they have a solid, rocky surface. In contrast, the outer planets are generally much larger and farther from the Sun and are mainly gaseous.
4. Answers may vary but should mention that Earth is one small inner planet in the solar system, which is part of a larger organization of stars in a galaxy, which is part of a larger organization of galaxies called the Milky Way. Answers should then mention that the Milky Way is in turn part of the universe, which contains all existing things.

CHAPTER 3

FORM A

Matching

1. i
2. e
3. k
4. c
5. j
6. f
7. g
8. a
9. l
10. h

Fill in the Blank

1. low-pressure
2. condensation
3. Tornadoes
4. Wind
5. temperatures
6. currents
7. middle-latitude
8. prevailing
9. marine west coast
10. low

Understanding Ideas

1. a
2. c
3. c
4. b
5. b

Practicing Skills

1. Pressure is affected by temperature which explains why the poles and equator have high and low pressure, respectively. The zones alternate because as the warm air rises it moves away from the equator and cools, causing it to sink again. A similar pattern is created as cold air moves away from the poles, warms, and rises.
2. westerlies in the Northern Hemisphere

Composing an Essay

1. Answers may vary but should mention that global winds are affected by Earth's temperature as well as its rotation.
2. Answers may vary but should mention that rotation, tilt, landforms such as mountain ranges, and ocean currents all affect climate.

FORM B

Short Answer

1. Forms of condensation include fog, clouds, dew, and frost. Forms of precipitation include rain, sleet, snow, and hail. Condensation remains in the air; precipitation falls out of it.
2. Low-pressure, humidity, temperature, and wind create storms.
3. A place's location on Earth determines its temperature, wind patterns, and air pressure, all of which are major factors in determining the climate.
4. The greenhouse effect causes heat energy to be trapped in Earth's atmosphere, raising world temperatures and leading to global warming.
5. The constant temperature of the ocean makes for milder climates in coastal areas. Ocean currents also move warm water to higher latitudes, creating climates that would not normally be found in those areas.

Practicing Skills

1. Earth's rotation causes the trade winds to blow toward the west as they flow toward the equator.
2. Cold air flowing away from the Poles meets warmer air at about 60° latitude and forces it upward. This rising air creates a low pressure zone.

Composing an Essay

1. Answers will vary but should mention that the wind belts affect air travel, shipping routes, and weather patterns (cyclones, tornadoes, and hurricanes) that

affect human life.

2. Answers will vary but might mention that parts of the western United States have desert climates because of the rain shadow on the leeward side of the western mountain ranges.
3. Most tropical climates are located in the low latitudes near equator region. The high temperatures, low air pressure, and lack of wind found there create the constant precipitation necessary for a tropical climate.
4. Answers will vary but should mention that the two tropical climates vary in the patterns of precipitation they receive. The tropical humid receives plentiful rainfall all year while the tropical wet and dry has distinct periods of very wet and very dry weather. The wet and dry climate also has a larger range of temperatures, which contributes to the longer dry periods.

Chapter 4

FORM A

Matching

1. d
2. e
3. a
4. b
5. i
6. j
7. h
8. f
9. k
10. g

Fill in the Blank

1. sediment
2. mantle
3. renewable
4. plate tectonics
5. headwaters
6. ore
7. An earthquake
8. erosion
9. topsoil or humus
10. oil

Understanding Ideas

1. a
2. a
3. a
4. c
5. b

Practicing Skills

1. 0 days
2. San Diego, CA

Composing an Essay

1. Answers may vary but should discuss the increased use of renewable and nonrenewable resources as well as steps that should be taken to conserve both types of resources. Students may mention that new ways to use Earth's resources could be developed.
2. Answers may vary but should mention that weathering alters the physical landscape and reduces humans' ability to grow crops in the soil.

FORM B

Short Answer

1. Wetlands provide a natural area for water runoff and they support large populations of fish, birds, plants, and shellfish.
2. The three types of plate movement are moving or spreading apart, colliding, and moving laterally.
3. The short-term effects of air pollution include smog and acid rain. Its long-term effects include serious damage to Earth's atmosphere through damage to the ozone layer and health problems for people.
4. The three layers of soil are topsoil, subsoil, and weathered rock fragments.
5. Three factors in the composition of soil are the type of rock the soil particles come from, the climate in which the rock is produced, and the type of landforms on which the soil is deposited.

Practicing Skills

1. approximately 40 days
2. Atlanta had the largest increase in the number of days with unacceptable air quality.

Composing an Essay

1. Answers will vary but should mention that people have tended to settle in flat areas where they can farm and where water is available. They have used rivers as transportation routes and have built railroads and highways along rivers.
2. Answers will vary but should mention that fossil fuels are nonrenewable resources that are becoming increasingly costly, and that the use of fossil fuels contributes to pollution of the environment.
3. Answers will vary but should mention that the hydrologic cycle is the movement of water through the hydrosphere. This

cycle is driven by solar energy, winds, and gravity. Water, when heated, turns to vapor. As it rises it cools and condenses, ultimately forming clouds that create precipitation. On land it can be stored in the ground or it can build up in plants or in bodies of water. Then the cycle begins again.

4. Answers will vary but should mention that soil exhaustion is a condition in which soil becomes nearly useless for farming. It can be prevented by using fertilizers to build up nutrients or by using crop rotation to prevent the exhausting of nutrients.

Chapter 5

FORM A

Matching

1. d
2. i
3. h
4. b
5. a
6. e
7. l
8. k
9. f
10. g

Fill in the Blank

1. birthrate
2. Africa
3. culture region
4. Migration
5. relocation
6. universalizing
7. traditionalism
8. Refugees
9. death rate
10. pull

Understanding Ideas

1. a
2. b
3. d
4. b
5. c

Practicing Skills

1. Stage 2
2. People who live in modern urban societies tend to marry later and have fewer children.

Composing an Essay

1. Answers may vary but should mention improved communication, transportation, and the Internet.
2. Answers may vary but should mention that overall the population density is much greater in Southeast Asia. There are high areas of population density within the United States, such as the northeast seaboard and the west coast, but the overall figure is very low. One reason is that the interior of the United States has several areas where living conditions are not ideal and most people choose not to live there. In contrast the entire area of Southeast Asia is heavily populated.

FORM B

Short Answer

1. birthrate, death rate, and migration
2. Cultures change through acculturation, innovation, and diffusion.
3. Globalization and traditionalism are the opposite of each other. Globalization occurs when connections around the world increase and cultures become more alike, while traditionalism involves following longtime practices and opposing many modern technologies and ideas.
4. English has become the lingua franca: most business and trade are conducted in English and English is the dominant language on the Internet.
5. Ethnic religions focus on one ethnic group and generally have not spread into other cultures. Old beliefs, legends, and customs are common and shape the religion. Animist religions often feature the belief in many gods as well as the belief in the presence of the spirits and forces of nature. Universalizing religions seek followers all over the world and appeal to many cultures. They are generally monotheistic.

Practicing Skills

1. The United States is in Stage 3 because both the birthrate and death rate are low. There is slow population growth.
2. Parents have many children, but poor health conditions result in a high infant mortality rate.

Composing an Essay

1. Answers will vary but should mention that in Stage 1 the birthrate and death rate are high which translates to little population growth. In Stage 2 the death rate drops but the birthrate stays high so population increases quickly. In Stage 3 both

the birthrate and the death rate are low so population growth is very slow.

2. Answers will vary but should mention that studying cultures allows us to better understand history, to predict population growth, to understand others better, and to forecast future developments.
3. Answers will vary but should mention that there were many pull factors for immigrants: free or cheap land, freedom from religious persecution, bountiful harvests, and good job opportunities. Migration to the United States has come in waves as push factors occurred in various countries (war, persecution, famine, and so on)
4. Answers will vary but should mention that culture regions often span more than one political boundary, which can cause political tension. Students may cite the Kurds living in Iraq, Turkey, and neighboring countries as an example. Students should also note that some political boundaries follow cultural divisions.

Chapter 6

FORM A

Matching

1. k
2. a
3. l
4. j
5. d
6. h
7. e
8. g
9. c
10. f

Fill in the Blank

1. capitalism
2. cities
3. infrastructure
4. market-oriented
5. Quaternary
6. cultural
7. United Nations
8. standard of living
9. democracies
10. subsistence

Understanding Ideas

1. d
2. a
3. c
4. d
5. b

Practicing Skills

1. The higher the percentage of people living in cities, the higher the number of television sets per 1,000 people. There are probably more televisions in cities because the telecommunications system is generally better in cities than in rural areas. Also, people living in cities probably have more disposable income.
2. Mali has only one doctor for every 18,376 people.

Composing an Essay

1. Answers may vary but should mention that developed countries tend to have market economies. Traditional and command economies are more common in developing countries.
2. Answers may vary but should mention that as a country develops, more people move to cities, which leads to changes in the rural geography. For example, farms may become larger and begin to grow crops for profit instead of subsistence.

FORM B

Short Answer

1. market economy, command economy, and traditional or subsistence economy.
2. A country's level of development can be measured by GNP, GDP, industrialization, and standard of living.
3. Cities have grown from basic cities with CBD to large cities with attendant edge cities. World cities have also developed.
4. Factors that can determine the location of a city include key resources, proximity to transportation and trade routes, and ease of defense.
5. Three geographic boundaries are natural (river, mountains), cultural (religious), and geometric (drawn by grid).

Practicing Skills

1. People living in developed countries have a life expectancy almost twice that of a person living in a developing country. People in developed countries have easier access to health care and other facilities that allow them to live longer.
2. There appears to be a direct correlation between literacy rate and GDP. If the GDP is high, so is literacy. If the GDP is low, so is the literacy rate.

Composing an Essay

1. Answers will vary but should mention that countries with high standards of living often depend on tertiary and quaternary economic activities, which require high levels of education.
2. Answers will vary but should mention that developed countries tend to have stable governments. Most have market economies. Poorer countries tend to have unstable governments and to depend on primary and secondary economic activities.
3. Answers will vary but should mention elements such as good government, benefits, education, improved infrastructure, and a market economy.
4. Answers will vary but should mention that geometric boundaries drawn without regard for cultural or ethnic divisions can cause wars or other problems, while countries that are divided along natural or cultural boundaries are often more unified. In addition, a country's boundaries can affect economic relationships. Landlocked countries, for example, might form trade agreements with coastal neighbors.

Unit 1

FORM A

Matching

1. j
2. i
3. a
4. c
5. d
6. l
7. h
8. g
9. f
10. k

Fill in the Blank

1. standard of living
2. physical
3. low
4. Migration
5. plate tectonics
6. elliptical
7. erosion
8. gravity
9. death rate
10. Earth's rotation

Understanding Ideas

1. a
2. b
3. d
4. a
5. c

Practicing Skills

1. 159° 58' E, 9° 25' S
2. 2,000 ft.

Composing an Essay

1. Answers will vary but should mention the six essential elements: the world in spatial terms, places and regions, physical systems, human systems, environment and society, and the uses of geography. These essential elements share many properties with the five themes but provide way of linking different themes together to make a more seamless study of geography.
2. Answers may vary but should mention improved communication, transportation, and the Internet.

FORM B

Short Answer

1. The four parts of the Earth system are the lithosphere, biosphere, hydrosphere, and atmosphere.
2. Three types of plate movement are moving or spreading apart, colliding, and moving laterally.
3. Three factors used by geographers to study changes in population are birthrate, death rate, and migration.
4. The three types of economic systems are market economy, traditional or subsistence economy, and command economy.
5. Geographers have superimposed a grid on the globe that includes lines of latitude and longitude. Lines of latitude run from east to west while lines of longitude run from north to south. These lines are measured in degrees that are further divided into minutes. The entire circle of the globe equals 360 degrees.

Practicing Skills

1. Guadalcanal appears to be an extremely mountainous island with a flat plain near sea level located on the north side of the island.
2. The sharpest drop in elevation occurs between Mt. Makarakomburu and the northern beach area, where the elevation drops from greater than 8,000 feet to sea level in a relatively short space.

Composing an Essay

1. Answers will vary but should mention elements such as good government, benefits, education, improved infrastructure, and a market economy.
2. Answers will vary but should mention that the two tropical climates vary in the patterns of precipitation they receive. The tropical humid receives plentiful rainfall all year while the tropical wet and dry has distinct periods of very wet and very dry weather. The wet and dry climate also has a larger range of temperatures, which contributes to the longer dry periods.
3. Answers will vary but should mention that the three types of map projections—conic, cylindrical, and flat-plane—serve different purposes. It is not possible to create a two-dimensional map that can accurately portray shape, size, and distance of a three-dimensional object such as Earth. Each type of map seeks to depict at least one aspect in an accurate fashion.
4. Answers will vary but should mention that people have tended to settle in flat areas where they can farm and where water is available. They have used rivers as transportation routes and have built railroads and highways along river valleys.

Chapter 7

FORM A

Matching

1. h
2. d
3. a
4. g
5. b
6. j
7. f
8. c
9. l
10. e

Fill in the Blank

1. rain shadow
2. north to south
3. Continental Divide
4. Alberta
5. ice cap
6. climate
7. newsprint
8. salmon
9. Arizona
10. hot spot

Understanding Ideas

1. a
2. d
3. c
4. a
5. c

Practicing Skills

1. C
2. F

Composing an Essay

1. Answers will vary. Settlers often established cities at the fall line because ships could sail upriver to that point, thus making these cities good places for trade and commerce. The many falls and fast flowing rivers at the fall line also allowed the building of waterwheels to power early industries.
2. Answers will vary but because North America has vast tracts of fertile soil, both countries are able to adequately feed their own populations as well as export the surplus. This is the first step a country must achieve to become developed. The vast forests, energy resources, and minerals found in North America have made both countries wealthy.

FORM B

Short Answer

1. Barrier islands are formed when ocean waves and currents deposit sand in shallow waters along a coastal plain.
2. The Canadian Shield is an arc of ancient rock that covers nearly half of Canada, stretching from the Arctic Ocean to the Atlantic Coast, centered on Hudson Bay.
3. Mountain ranges west of the Great Plains include the Sierra Nevada, the Cascade Range, the Brooks Range, and the Rocky Mountains.
4. High mountains block prevailing winds, creating a rain-shadow effect on the intermountain region. This rain-shadow brings plentiful rain to the windward sides of the mountains and creates arid and semi-arid climates on the leeward sides of the mountains. The high elevation of the mountain ranges means that some high elevations have a highland climate.
5. The two main climates of Canada are tundra and subarctic.

Practicing Skills

1. A and D
2. B

Composing an Essay

1. The Mississippi, Missouri, and Ohio rivers and their tributaries form the major river systems east of the Continental Divide. The St. Lawrence River system connects the Great Lakes to the Atlantic Ocean and drains most of southeastern Canada. The Mackenzie River system drains the interior plains and part of the Canadian Shield. The Columbia and Fraser rivers flow into the Pacific Ocean and the Colorado River flows southwestward into the Gulf of California.
2. Answers will vary but students should mention that the Appalachians were at one time a barrier to westward expansion. The Sierra Nevada and Rocky Mountains in the West also posed a barrier to movement. Populations are denser on the eastern side of the Appalachians and western side of the Rockies and Sierra Nevada partly because these ranges made movement toward the interior difficult for early settlers.
3. The Gulf Stream is an ocean current that brings warm tropical water northward along eastern North America. The warm water makes temperatures in the region more consistent.
4. They differ in that the landforms of eastern North America are much older and more eroded than those in the west.

Chapter 8

FORM A

Matching

1. j
2. e
3. i
4. h
5. l
6. a
7. k
8. c
9. d
10. b

Fill in the Blank

1. Spanish
2. earthquakes
3. Asia; Latin America
4. Asia
5. Dallas-Ft. Worth
6. Corn
7. southern
8. fertilizers
9. Northeast
10. South

Understanding Ideas

1. b
2. a
3. b
4. c
5. b

Practicing Skills

1. 1900
2. World War II, the Holocaust

Composing an Essay

1. Answers will vary but should mention that the contact of many different cultural groups led to the creation of new forms of music like jazz, country, and rap. Architectural styles such as skyscrapers reflect the huge growth of the population. American literature reflects the country's rich cultural and artistic heritage.
2. Answers may vary but should mention that people settle where there are jobs. For example, densely populated areas include urban regions that house industry and a wide range of services. Farmlands that do not require large numbers of workers are more sparsely populated. Desert areas that have little industry or good farmland are the most sparsely populated.

FORM B

Short Answer

1. They wanted power to be distributed among the local, state, and national levels, with the ultimate power resting with the people.
2. Possible Indian names include Alaska, Arizona, Kansas, Nebraska, North and South Dakota, Mississippi, and Appalachians. Possible French names include New Orleans, St. Louis, Dubuque, and Detroit. Possible British names include New York, New Jersey, New England, New Hampshire, Georgia, Virginia, Richmond, and Charleston.
3. Ohio, Indiana, Illinois, Michigan, Wisconsin, Minnesota, Iowa, and Missouri
4. Possible answers include acid rain, overuse of fertilizers, limited water resources, population and urban growth issues, and issues associated with economic development.
5. Students might note that the country

fought two world wars, experienced a Great Depression, and worked through a Cold War with Russia to emerge as a strong superpower at the end of the 1900s.

Practicing Skills

1. 1900 and 1960
2. Answers will vary but should note that the foreign-born population will likely increase.

Composing an Essay

1. Answers will vary but students should mention that the middle-latitude location of the United States and its favorable climates have allowed the country to become a large producer of food. The diversity of the population has fostered the diffusion of ideas and innovations. Rich mineral and energy resources have allowed the country to grow strong as well. A democratic government has allowed the country to have a mostly peaceful, stable economy and social fabric.
2. Answers will vary but should mention that historically most immigrants were European and settled in the Northeast. Now most immigrants are from Asia and Latin America and settle in many different regions.
3. Answers will vary but should mention that cheap land, favorable laws and regulations, a warm climate, and lower wages have attracted many new industries to the South.
4. Answers will vary but should mention that southern colonies focused on growing tobacco and cotton, while northern colonies focused on ship building, fishing, and trade.

Chapter 9

FORM A

Matching

1. a
2. i
3. k
4. g
5. b
6. h
7. f
8. l
9. c
10. e

Fill in the Blank

1. manufacturing
2. NAFTA
3. Atlantic
4. Atlantic
5. Vikings
6. regionalism
7. Nunavut
8. market
9. Vancouver
10. Jacques Cartier

Understanding Ideas

1. c
2. b
3. a
4. c
5. b

Practicing Skills

1. Northwest Territories
2. harsh climate, small population, remote location

Composing an Essay

1. Answers will vary but should include reasoning to support students' opinions.
2. Answers will vary but each viewpoint should be supported with facts from the text.

FORM B

Short Answer

1. France and Great Britain have been the most influential countries in Canada's development.
2. These areas are the most densely settled and economically developed.
3. The cities in the Prairie Provinces have grown due to their proximity to oil resources and agricultural lands, and their placement along an important rail route.
4. Nunavut is Canada's newest territory. It was created out of the Northwest Territories in 1999 to give the Inuit of the region a self-governing homeland.
5. Quebec has a strong French-Canadian culture, French is widely spoken there, and most people who live in Quebec are Roman Catholic.

Practicing Skills

1. Northwest Territories
2. British Columbia

Composing an Essay

1. Answers will vary but should mention that English is the main language in the

United States and most of Canada; both countries share similar histories as former British colonies; both have welcomed many immigrants; the two countries are tied economically.

2. Answers will vary but should mention that the southern areas of Ontario and Quebec are the most densely populated because it was in this region that the Europeans first settled and because this region has the most favorable climate and geography for human habitation.
3. Answers will vary but should mention that the eastern provinces would be separated from the rest of Canada. A large portion of the population would be gone along with some of the major cities including the major seaport of Montreal.
4. Answers will vary but should mention that it has been Canada's poorest region, with the lowest wages and the highest unemployment. Also, long cold winters and thin rocky soils make farming difficult there.

Unit 2

FORM A

Matching

1. a
2. f
3. i
4. k
5. e
6. l
7. g
8. h
9. d
10. c

Fill in the Blank

1. NAFTA
2. South
3. orographic effect
4. Asia; Latin America
5. regionalism
6. Arizona
7. Corn
8. Continental Divide
9. Nunavut
10. Vancouver

Understanding Ideas

1. c
2. c
3. c
4. b
5. d

Practicing Skills

1. Northwest Territories
2. harsh climate, small population, remote location

Composing an Essay

1. Answers will vary but each viewpoint should be supported by facts from the text.
2. Answers will vary but should mention that the contact of many different cultural groups led to the creation of new forms of music like jazz, country, and rap. Architectural styles such as skyscrapers reflect the huge growth of the population. American literature reflects the country's rich cultural and artistic heritage.

FORM B

Short Answer

1. Nunavut is Canada's newest territory. It was created out of the Northwest Territories in 1999 to give the Inuit of the region a self-governing homeland.
2. Mountain ranges west of the Great Plains include the Sierra Nevada, the Cascade Range, the Brooks Range, and the Rocky Mountains.
3. They wanted power distributed among the local, state, and national levels, with the ultimate power resting with the people.
4. The cities in the Prairie Provinces have grown due to their proximity to oil resources and agricultural lands, and their placement along an important rail route.
5. Possible answers include acid rain, overuse of fertilizers, limited water resources, population and urban growth issues, and issues associated with economic development.

Practicing Skills

1. Northwest Territories
2. British Columbia

Composing an Essay

1. Answers will vary but students should mention that the middle-latitude location of the United States and its favorable climates have allowed the country to become a large producer of food. The diversity of the population has fostered diffusion of ideas and innovations. Rich mineral and energy resources have allowed the country to grow strong as well. A democratic

government has allowed the country to have a mostly peaceful, stable economy and social fabric.

2. Answers will vary but should mention that the eastern provinces would be separated from the rest of Canada. A large portion of the population would be gone, along with some of the major cities. The major seaport of Montreal would be lost.
3. They differ in that the landforms of eastern North America are much older and more eroded than those in the west.
4. Answers will vary but should mention that English is the main language in the United States and most of Canada; both countries share similar histories as former European colonies; both have welcomed many immigrants; the two countries are tied economically.

Chapter 10

FORM A

Matching

1. g
2. l
3. e
4. a
5. c
6. i
7. f
8. h
9. b
10. j

Fill in the Blank

1. infrastructure
2. land
3. Milpa
4. limestone
5. Isthmus of Tehuantepec
6. Sierra Madre del Sur
7. Mestizos
8. mountains
9. smallpox
10. Southern Mexico

Understanding Ideas

1. b
2. a
3. a
4. d
5. d

Practicing Skills

1. Canada, with 82 percent
2. Canada

Composing an Essay

1. Answers may vary but should mention that when the conquistadores arrived, they brought smallpox and other diseases unknown to the region. The Indians had no resistance to these diseases and died in large numbers. The Spanish also introduced horses and muskets and forged alliances with the enemies of the Aztecs.
2. Answers may vary but should mention that Mexico has shifted from a rural agricultural country to a mostly urban and industrial country. Family size has decreased. Traffic, crime, and all the other attendant problems of industrialized countries exist here. Mexico's government has become increasingly democratic. It is a strong partner in NAFTA.

FORM B

Short Answer

1. A limestone base and erosion create the optimal conditions for sinkholes.
2. Sierra Madre del Sur, Sierra Madre Occidental, and Sierra Madre Oriental
3. cement, construction, plastics, textiles, tourism, and chemicals
4. Aztec, Toltec, Zapotec, Maya, and Olmec
5. *Ejidos* were lands owned by the Indians and worked as a group. Plantation workers did not own the land they worked.

Practicing Skills

1. Canada, with a difference of $16.7 billion
2. $15.8 billion

Composing an Essay

1. Answers will vary but should mention that towns often began as missions. Towns grew around a church or mission, the plaza in front of the church becoming the center of village life and community markets.
2. Answers will vary but should mention that throughout Mexico's history, the Spanish and Indian cultures mixed. Early in the colonial period, most of the colonists were men, and marriage with American Indian women was common. Hence, most Mexicans today are of mixed ancestry. Such mixing did not occur as often in British or French colonies.
3. Answers will vary but should mention economic inequality, crime, and poor infrastructure.
4. Answers will vary but should mention the three major mountain ranges and the

Mexican Plateau as the major landforms of Mexico. Mexico's climate varies widely due to its closeness to the equator, which creates the tropical climates, and its mountains, which create highland climates and rain shadows.

Chapter 11

FORM A

Matching

1. a
2. e
3. j
4. c
5. l
6. f
7. i
8. d
9. h
10. b

Fill in the Blank

1. the former Soviet Union
2. Honduras
3. sugarcane
4. warm and sunny climate
5. Haiti and Jamaica
6. Belize
7. Costa Rica
8. Guatemala
9. United States
10. tectonic processes

Understanding Ideas

1. c
2. a
3. c
4. b
5. d

Practicing Skills

1. about 10 inches
2. September

Composing an Essay

1. Answers will vary but should note that unlike the other countries of the Caribbean, Cuba has a command economy—its Communist government makes all of the decisions about production. Also, Cuba, unlike the other countries, is not a member of CARICOM, the region's economic union. Cuba's former dependence on the Soviet Union as a market partner has also been detrimental since the collapse of the Soviet Union.
2. Answers will vary but should mention that ecotourism allows the country to preserve the nature reserves of the country while at the same time providing a substantial boost to the economy. The arrival of additional tourists also spurs growth in other related industries.

FORM B

Short Answer

1. The meeting of tectonic plates—the Caribbean Plate with the North and South American Plates—has created many mountains in the area. The tops of these volcanic mountains are islands in the Lesser Antilles.
2. Costa Rica has a tradition of democracy, education, and political stability. These factors have attracted a great deal of foreign investment. Costa Rica has also been a leader in developing tourism.
3. Christopher Columbus landed on an island in the southern Bahamas in 1492, but he thought he had reached islands off the coast of Asia that Europeans called the Indies. Even though Columbus was wrong, the term West Indies stuck.
4. The United States built the Panama Canal in the early 1900s. It has been important because it allows ships to move quickly from the Atlantic Ocean and Caribbean Sea to the Pacific Ocean.
5. Creole is a blend of European, African, or Caribbean Indian languages. Papiamento is a creole language that combines elements of Spanish, Dutch, and Portuguese.

Practicing Skills

1. January, February
2. approximately 69 inches

Composing an Essay

1. Answers will vary but should mention the diversity of languages spoken (European languages, creole languages, and Papiamento) and the variety of religious beliefs (Roman Catholicism, Hinduism, voodoo, and Santeria).
2. Answers will vary but should mention that wealth is still concentrated in the hands of a small number of families; Christianity, particularly the Roman Catholic Church, remains important; and Spanish is the official language of all of the region's countries except Belize, and English is the official language there.
3. Answers will vary but should mention

that rapid population growth has resulted in high rates of unemployment and underemployment in the region.

4. Answers will vary but should mention that Puerto Rico is a commonwealth of the United States. Puerto Ricans are U.S. citizens, but Puerto Rico has no voting representatives in the U.S. Congress.

Chapter 12

FORM A

Matching

1. c
2. f
3. h
4. a
5. g
6. i
7. j
8. k
9. l
10. e

Fill in the Blank

1. Chile
2. tropical humid
3. Bolivia
4. copper
5. Paraguay
6. Maracaibo
7. Colombia
8. Francisco Pizarro
9. Brazil
10. Chibcha

Understanding Ideas

1. c
2. b
3. c
4. b
5. d

Practicing Skills

1. Northeast
2. Southeast

Composing an Essay

1. Answers will vary but should note that Brazilian independence followed a different path because the ruling Portuguese royal family came to live in the colony. The country only received its independence after the Portuguese king returned to Portugal. In contrast, the Spanish colonies had gained their independence decades earlier after the authorities fled the country after unrest.
2. Answers will vary but should mention that the Amazon River is the world's largest river in volume and that it drains the largest area in the world.

FORM B

Short Answer

1. The collision of the Nazca and South American plates created the Andes. Continued tectonic activity causes volcanic eruptions and earthquakes.
2. South America has a wide variety of climate regions because the continent extends across 60 degrees of latitude.
3. Patagonia lies in the rain shadow of the Andes, which creates semiarid and arid climates.
4. Answers will vary. Students may discuss how violence is affecting voting patterns, civil rights, politics, and justice in Colombia. Students should attribute the violence at least in part to the illegal drug trade in that country.
5. suspension bridges, stone construction, terraced fields, paved roads

Practicing Skills

1. East Central and South
2. It is located in the Amazon River basin, which is a tropical rain forest.

Composing an Essay

1. Answers will vary but should mention that millions of South American Indians died from diseases brought by the Europeans; many more died in battles with the Europeans; and others died from overwork in Spanish mines and on ranches and plantations.
2. Answers will vary but should mention that densely populated areas of South America hug the coasts and reach only a few hundred miles inland. Many of the major cities are seaports. The interior of the continent is in general thinly populated. Each country tends to have one or two large cities, which are much larger than any other cities in the country.
3. Answers will vary but should mention that the independence movements did little to improve most people's lives. Although revolutions often changed the governments in the new countries, they only tended to replace one group of powerful families with another.
4. Answers will vary but should mention that

the purpose of Mercosur is to expand trade, improve transportation, and reduce tariffs among member countries. Argentina, Brazil, Paraguay, and Uruguay are full members of this trade organization.

Unit 3

FORM A

Matching

1. a
2. k
3. i
4. e
5. g
6. f
7. l
8. j
9. h
10. d

Fill in the Blank

1. Belize
2. Isthmus of Tehuantepec
3. copper
4. Chibcha
5. land
6. warm and sunny climate
7. infrastructure
8. United States
9. tropical humid
10. Milpa

Understanding Ideas

1. c
2. b
3. c
4. b
5. a

Practicing Skills

1. South and East Central
2. Northeast

Composing an Essay

1. Answers may vary but should mention that Mexico has shifted from a rural agricultural country to a mostly urban and industrial country. Family size has decreased. Traffic, crime, and all the other attendant problems of industrialized countries exist here. Mexico's government has become increasingly democratic. It is a strong partner in NAFTA.
2. Answers will vary but should mention that ecotourism allows the country to preserve the nature reserves of the country while at the same time providing a substantial boost to the economy. The arrival of additional tourists also spurs growth in other related industries.

FORM B

Short Answer

1. South America has a wide variety of climate regions because the continent extends across 60 degrees of latitude.
2. The meeting of tectonic plates—the Caribbean Plate with the North and South American Plates—has created many mountains in the area. The tops of these volcanic mountains are islands in the Lesser Antilles.
3. Answers will vary. Students may discuss how violence is affecting voting patterns, civil rights, politics, and justice in Colombia. Students should attribute the violence at least in part to the illegal drug trade in that country.
4. A limestone base and erosion create the optimal conditions for sinkholes.
5. Christopher Columbus landed on an island in the southern Bahamas in 1492, but he thought he had reached islands off the coast of Asia that Europeans called the Indies. Even though Columbus was wrong, the term West Indies stuck.

Practicing Skills

1. January, February
2. approximately 69 inches

Composing an Essay

1. Answers will vary but should mention the three major mountain ranges and the Mexican Plateau as the major landforms of Mexico. Mexico's climate varies widely due to its closeness to the equator, which creates tropical climates, and its mountains create the highland climates and rain shadows.
2. Answers will vary but should mention that rapid population growth has resulted in high rates of unemployment and underemployment in the region.
3. Answers will vary but should mention that towns often began as missions. Towns grew around a church or mission, the plaza in front of the church becoming the center of village life and community markets.
4. Answers will vary but should mention that the independence movements did little to improve most people's lives.

Although revolutions often changed the governments in the new countries, they only tended to replace one group of powerful families with another.

Chapter 13

FORM A

Matching

1. k
2. c
3. h
4. b
5. j
6. e
7. d
8. f
9. a
10. g

Fill in the Blank

1. navigable rivers
2. North Atlantic Drift
3. North
4. Loess
5. Northern European Plain
6. coal
7. mineral
8. farming
9. Alpine Mountain system
10. temperate forest

Understanding Ideas

1. a
2. a
3. b
4. b
5. a

Practicing Skills

1. Spain
2. Spain

Composing an Essay

1. Answers will vary but should mention that the North Atlantic Drift, a warm ocean current, causes the mild temperatures. The North Atlantic Drift warms the air above it, and prevailing westerly winds carry this warm moist air across much of northwestern Europe.
2. Answers will vary but should mention that Europe has become increasingly dependent on imported oil. Oil and gas from Southwest Asia, Russia, and Africa power the economies of most European countries.

FORM B

Short Answer

1. Glaciers scoured the landscapes of Scandinavia and much of the British Isles, carved fjords along Norway's coast, and left behind thin soils and thousands of lakes.
2. Europe's three major climate types are the marine west coast climate, the humid continental climate, and the Mediterranean climate.
3. The major landforms within the Central Uplands include the Massif Central, the Jura Mountains, and the Bohemian Highlands.
4. The Po River Valley and the Guadalquivir River Valley are important farming areas.
5. Trees such as ash, beech, maple, and oak are common in the temperate forest biome, as are badgers, deer, and a variety of birds.

Practicing Skills

1. Norway and Iceland
2. Norway

Composing an Essay

1. Answers will vary but should mention that for thousands of years Europeans have hunted animals and cleared forests for timber and farmland. The growth of towns, cities, and roads has also changed the natural environment. Some waterways have been polluted. Many species have become extinct from the resulting loss of habitat.
2. Answers will vary but should mention that the North Atlantic Drift moderates the climates of the region by warming the air above it. This warm air, carried by the prevailing winds over much of Europe, brings mild temperatures and rain and creates conditions that allow crops to grow much farther north than normal and for seaports in northern regions to stay free of ice.
3. Answers will vary but should mention that throughout history fishing has been an important part of Europe's economy. Fishing villages dot Europe's coasts, and fishing boats can be found in all waters bordering Europe.
4. Answers will vary but should mention that Europe is nearly surrounded by water. Its long coastline has hundreds of natural

harbors, which are often located near the mouths of navigable rivers.

CHAPTER 14

FORM A

Matching

1. h
2. j
3. e
4. d
5. l
6. b
7. c
8. f
9. g
10. k

Fill in the Blank

1. Algeria and Morocco
2. Lutheran
3. France
4. Denmark
5. Angles and Saxons
6. language; religion
7. geysers
8. Luxembourg
9. potato famine
10. Flanders; Wallonia

Understanding Ideas

1. a
2. a
3. d
4. d
5. c

Practicing Skills

1. on the Île de la Cité
2. Arc de Triomphe

Composing an Essay

1. Answers will vary but students should note that many countries have tried nationalizing and then privatizing industries. In addition, they have established extensive welfare systems that have resulted in high standards of education and health care. A mix of capitalism and socialism is prevalent in many of the Scandinavian countries. All of them have strong, diversified economies and healthy populations.
2. Answers will vary but should note that the violence is caused by a dispute over Northern Ireland. The majority of Northern Ireland's population is descended from Protestant English who want to remain part of Great Britain. However, the large minority group of Roman Catholics wants to return the area to the Republic of Ireland where they would be part of the majority.

FORM B

Short Answer

1. Answers may include the Celts, Romans, Angles, Saxons, Vikings, or Normans.
2. The economy of Ireland has shifted from a largely agricultural base to one that is more industrial.
3. Many peoples from overseas departments have immigrated to France looking for jobs. Immigration from Algeria and Morocco is particularly large. These diverse groups are introducing their own cultures to France.
4. Denmark, Iceland, Sweden, Finland, and Norway
5. They are all small countries sandwiched between two larger countries—Germany and France. They also share many historical and cultural features.

Practicing Skills

1. Champs Élysées
2. Luxembourg Gardens

Composing an Essay

1. Answers will vary but should mention that the two countries are similar in that their populations speak English and exhibit many Celtic influences. Proximity to each other has meant that they share many culture traits. They are different in their forms of government, Ireland being a republic and Great Britain a constitutional monarchy. They also differ in their religions, Ireland being predominantly Roman Catholic while Great Britain is mainly Protestant.
2. Answers will vary but should mention that France has a highly diversified economy. It has a very strong agricultural base, and supports large fashion, perfume, cosmetics, glassware, and furniture industries. Tourism is also a large part of the economy. France's once large sector of heavy industry is in decline but high-tech industries are growing in importance.
3. Answers will vary but should mention that most Scandinavians live in the southern parts of their countries where climates are warmer. In addition, most people live along

the coast and in or near their capital cities.

4. Answers will vary but should mention that being at the center of the Industrial Revolution gave Great Britain economic and political advantages over most other countries. The economy grew as British manufacturers produced goods that were desirable everywhere. This paved the way for the British to establish colonies, which has allowed British culture to spread throughout the world and has established English as the lingua franca in many parts of the world.

Chapter 15

FORM A

Matching

1. d
2. k
3. j
4. e
5. i
6. h
7. f
8. c
9. l
10. b

Fill in the Blank

1. Hungary
2. Gdańsk
3. Charlemagne
4. Switzerland
5. Finns
6. Germany
7. Bratislava
8. Bern
9. Austria
10. Velvet Revolution

Understanding Ideas

1. a
2. c
3. c
4. a
5. d

Practicing Skills

1. The kingdom of Hungary was much larger.
2. Poland

Composing an Essay

1. Answers will vary but should note that Poland began building a market economy after the collapse of the Soviet Union. The country has attracted foreign investment, and Polish companies are growing. This economic growth was furthered by the adoption of a new constitution that promised to continue to turn over many government-controlled companies to private ownership.

2. Answers will vary but should mention that Germany was forced to accept harsh peace terms after World War I. Germany's economy then collapsed in the 1920s. Food shortages, high inflation, and high unemployment brought about severe hardships. These problems helped bring the Nazi Party to power.

FORM B

Short Answer

1. Vienna was Central Europe's political and cultural center under the Habsburgs and remains Austria's capital and largest city. It has historic palaces, churches, and performance halls. In addition, many great artists and composers once lived there.

2. The Baltic countries are Latvia, Lithuania, and Estonia. Trade is essential to these countries because they have small populations and limited natural resources.

3. Migration from rural areas to Budapest slowed down as rural areas experienced improvements in water, sewage, and other services. Also, many people have been moving out of Budapest to live in the suburbs.

4. Switzerland has four major languages, and no religion holds an overwhelming majority among the population.

5. Much of Poland's farming activity takes place on land made fertile by thick deposits of loess. Polish farmers grow cereals, potatoes, and sugar beets.

Practicing Skills

1. Germany and Russia
2. Austria, Czechoslovakia, Hungary, Poland, Yugoslavia

Composing an Essay

1. Answers will vary but should include the following information: German is the dominant language, although there are several regional dialects. German writers have contributed significantly to world literature. Most Germans are either Roman Catholic or Protestant, though the southern and western portions are more Catholic than the northern and eastern. Many Germans do not attend religious services of any kind. German food features pork, sausages, veal, and cheeses. Rich pastries are popular.

2. Answers will vary but should mention that Kaliningrad, then called Konigsberg, was once part of Germany. During WWII it became a staging area for German attacks on the Soviet Union. When the Soviets defeated Germany they took control of the area. Large numbers of Russians moved there, and Kaliningrad became a key Soviet naval base. It remained under Russian control even after the collapse of the Soviet Union.
3. Answers will vary but should mention that the Czech Republic has good supplies of coal, iron ore, and uranium and consequently is known for the production of fine steel and glass products. It has one of the strongest and most stable economies of the former Soviet Union. Slovakia is the poorer of the two countries. Slovakia's move to a capitalist system has been difficult with high unemployment.
4. Answers will vary but should mention that Hungarians speak Magyar and are predominantly Roman Catholic.

Chapter 16

FORM A

Matching

1. b
2. f
3. a
4. l
5. c
6. e
7. g
8. h
9. k
10. j

Fill in the Blank

1. Italy
2. Francisco Franco
3. Greece
4. Benito Mussolini
5. Croatia
6. Madeira, the Azores
7. Moors
8. Andorra
9. Basque
10. Porto

Understanding Ideas

1. b
2. c
3. c
4. d
5. b

Practicing Skills

1. the 30–34 age group
2. the 80+ age group

Composing an Essay

1. Answers may vary but should mention that the eastern Balkans are more closely tied to Russia, as they use the Cyrillic alphabet and follow Eastern Orthodox Christian religions. They were more firmly under the control of Russia after World War II. The eastern Balkans are more politically and economically stable and have not experienced the ethnic conflicts that have affected the western Balkans. The western Balkans are much more ethnically diverse. Their economies have suffered greatly from conflict. The two regions are alike in that they are both very poor and they both were under the influence of Communist regimes during most of the second half of the 1900s.
2. Answers may vary but should mention the invasion of the Moors in 700 A.D. and their contributions to Spanish culture before they were pushed out in 1492. Spain's naval dominance in the 1500s allowed the country to establish vast colonies in the Americas. By the end of the 1800s Spain had lost nearly all its colonies. The 1900s were marked by a terrible civil war that began in 1936 followed by years of rule by a dictator. Upon his death in 1975, Spain became a democratic country with a constitutional monarchy.

FORM B

Short Answer

1. Portugal's history mirrors Spain's in the fact that it too was ruled by the Romans and the Moors. Influences of these empires are evident in the countries' languages. Portugal too had a successful period of colonization that is now much diminished. Both countries now have democratic governments and immigration problems. Tourism is a major source of income for both countries.
2. Albania, Bosnia and Herzegovina, Croatia, Macedonia, Slovenia, and Serbia and Montenegro

3. Greece is struggling as one of the poorest countries in Europe. Illegal immigration, low population growth, and rapid urbanization with its attendant pollution are its chief concerns.
4. The great success of trading cities in Italy provided the additional money necessary to sponsor artists and sculptors. Trade also allowed the ideas of the Renaissance to be spread throughout the region.
5. Physical isolation and international treaties protect the microstates. Tourism and trade have helped them stay independent. Low taxes make them attractive to investors and tourists.

Practicing Skills

1. It has decreased.
2. the 80+ age group

Composing an Essay

1. Answers will vary but should mention that the grouping of several religious and ethnically distinct groups under one communist regime and then the subsequent disbanding of the government put the groups in conflict.
2. Answers will vary but should mention that the Balkans were influenced by three very different empires (Ottoman, Roman, and Russian) while other areas were not occupied by so many peoples. These empires brought their own cultures and religions to the Balkans. The area is also geographically at a crossroads of Asia, Europe, and Russia, ensuring that many peoples have passed through the area.
3. Answers will vary but should mention that Spanish is now spoken by over 400 million people around the world due to Spain's colonial history. South and Central America also reflect much of the culture of Spain in the way cities are built and organized.
4. Answers will vary but should mention that Italy has brought its language, art, architecture, and religion to much of the world. Many of the languages of Europe are derived from Latin, and Roman architecture is visible throughout Europe.

Unit 4

FORM A

Matching

1. f
2. a
3. g
4. h
5. c
6. d
7. j
8. b
9. e
10. k

Fill in the Blank

1. Charlemagne
2. France
3. North
4. Benito Mussolini
5. Flanders; Wallonia
6. Germany
7. temperate forest
8. Croatia
9. agricultural
10. Andorra

Understanding Ideas

1. a
2. c
3. a
4. c
5. b

Practicing Skills

1. on the Île de la Cité
2. Arc de Triomphe

Composing an Essay

1. Answers will vary but should mention that the mild temperatures are caused by the North Atlantic Drift, a warm ocean current. The North Atlantic Drift warms the air above it, and prevailing westerly winds carry this warm moist air across much of northwestern Europe.
2. Answers will vary but should note that Poland began building a market economy after the collapse of the Soviet Union. The country has attracted foreign investment, and Polish companies are growing. This economic growth was furthered by the adoption of a new constitution that promised to continue to turn over many government-controlled companies to private ownership.

FORM B

Short Answer

1. The Baltic countries are Latvia, Lithuania, and Estonia. Trade is essential to these countries because they have small

populations and limited natural resources.

2. Greece is struggling as one of the poorest countries in Europe. Illegal immigration, low population growth, and rapid urbanization with its attendant pollution are its chief concerns.
3. Denmark, Iceland, Sweden, Finland, and Norway
4. The great success of trading cities in Italy provided the additional money necessary to sponsor artists and sculptors. Trade also allowed the ideas of the Renaissance to be spread throughout the region.
5. Glaciers scoured the landscapes of Scandinavia and much of the British Isles, carved fjords along Norway's coast, and left behind thin soils and thousands of lakes.

Practicing Skills

1. Champs Élysées
2. Luxembourg Gardens

Composing an Essay

1. Answers will vary but should mention that the Balkans were influenced by three very different empires (Ottoman, Roman, and Russian) while other areas were not occupied by so many peoples. These empires brought their own cultures and religions to the Balkans. The area is also geographically at a crossroads of Asia, Europe, and Russia, ensuring that many peoples have passed through the area.
2. Answers will vary but should mention that being at the center of the Industrial Revolution gave Great Britain economic and political advantages over most other countries. The economy grew as British manufacturers produced goods that were desirable everywhere. This paved the way for the British to establish colonies, which has allowed British culture to spread throughout the world and has established English as the lingua franca in many parts of the world.
3. Answers will vary but should mention that Europe is nearly surrounded by water. Its long coastline has hundreds of natural harbors, which are often located near the mouths of navigable rivers.
4. Answers will vary but should mention that the Czech Republic has good supplies of coal, iron ore, and uranium and consequently is known for the production of fine steel and glass products. It has one of the strongest and most stable economies of the former Soviet Union. Slovakia is the poorer of the two countries. Slovakia's move to a capitalist system has been difficult with high unemployment.

Chapter 17

FORM A

Matching

1. f
2. g
3. j
4. h
5. b
6. c
7. i
8. e
9. k
10. l

Fill in the Blank

1. Moscow
2. Cossacks
3. Siberia
4. Sakhalin Island
5. Eastern Orthodox Christianity
6. Vladimir Lenin
7. Donets Basin
8. Autarky
9. steppe
10. Ivan IV (the Terrible)

Understanding Ideas

1. a
2. c
3. d
4. b
5. a

Practicing Skills

1. D
2. F

Composing an Essay

1. Answers will vary but should mention that autarky is a system in which a country tries to produce all of the goods it needs. A communist country might want to follow such a policy in order to limit its association with capitalist countries.
2. Answers may vary but should mention that Moscow is the national center of communications, culture, education, finance, politics, and transportation. Many institutions of higher learning are found there. Roads, rails, and air routes link

Moscow to all points in Russia.

FORM B

Short Answer

1. Russian traditions that have survived changes in government include religious preferences, an emphasis on education, food and drink choices, and the popularity of dachas.
2. Much of Russia's rich forest, energy, and mineral resource wealth has been wasted because the government stressed production over conservation. Some of the resources that remain are in remote areas or are of low quality.
3. The mountain ranges include the Ural Mountains, the Carpathian Mountains, and the Caucasus Mountains.
4. Russia has large deposits of oil, coal, and natural gas. Russia also has substantial mineral and lumber resources.
5. The majority of Russians are Eastern Orthodox Christians. There are also Roman Catholic populations found along the western borders near Poland and the Czech Republic. Muslim populations are located in the south. Protestant Christianity is gaining popularity due to missionary projects.

Practicing Skills

1. H
2. B

Composing an Essay

1. Students may say that it was unsuccessful. Without competition from capitalist countries, industrial efficiency and product quality fell. Production of consumer goods and services lagged behind those of capitalist countries. In addition, agricultural production suffered.
2. Answers will vary but should mention that changes in government have been relatively peaceful after free elections. However, crime and unemployment have increased, making people fearful. The health-care system has disintegrated, leaving many elderly with no safety net. Many of those who are getting rich are trying to avoid paying taxes. The middle-classes do not feel secure, fearing that the government may take over business and industry again.
3. Answers will vary but should mention that rich oil and gas fields lie between the Volga River and the Ural Mountains. These fields have been crucial to the region's development.
4. Answers will vary but should mention that the countries of the region are working to develop light industry, which focuses on the production of consumer goods. Heavy industry is becoming less important.

Chapter 18

FORM A

Matching

1. f
2. j
3. a
4. l
5. k
6. i
7. d
8. c
9. g
10. b

Fill in the Blank

1. Aral Sea
2. water
3. Islam
4. Mediterranean
5. Kyrgyzstan
6. Timur
7. longer
8. Kopet-Dag
9. cotton
10. Turkic

Understanding Ideas

1. b
2. b
3. b
4. c
5. a

Practicing Skills

1. Tajikistan
2. It is geographically closer to Russia than are the other countries.

Composing an Essay

1. Answers will vary but should mention political problems and ethnic conflict. Conflicts over water resources must also be resolved. The countries must continue to clean up the results of Soviet agricultural, industrial, and military practices that damaged the land and water.
2. Answers may vary but should mention that the Soviets closed more than 35,000 mosques, churches, and Islamic schools. Many of these building were abandoned

or destroyed over time. Since 1991 the remaining buildings have reopened.

FORM B
Short Answer
1. Central Asia has relatively few big cities because, throughout most of its history, most of the region's people have been nomads or farmers.
2. Altay Shan, Pamirs, Tian Shan, Kopet-Dag
3. Timur supported the arts, literature, and science. He built many gardens, mosques, and palaces in Samarqand.
4. Deer, pheasant, wild boar, and snow leopards are all found in the mountain ranges of Central Asia.
5. The Russians built railroads and irrigation projects and expanded cotton production and oil exports. The railroads offered better access to the region's resources.

Practicing Skills
1. Uzbekistan
2. Tajikistan

Composing an Essay
1. Answers will vary but should mention that railroads helped the Russians create a stronger military presence between India—which was under the control of the British—and Russia. The railroads also gave the Russians better access to Central Asia's resources.
2. Answers will vary but should mention that the Soviets replaced Islamic schools with state-run schools and made public services widely available. Today, literacy rates in the region are well above world averages and life expectancy in most of the region's countries is longer than the world average, particularly for women.
3. Answers will vary but should mention that major irrigation rivers and canals cross international boundaries, forcing countries to share resources. Exclaves of one country within another make boundaries difficult to control, a situation that has led to an increase in illegal drug smuggling.
4. Answers will vary but should mention the following factors: outdated equipment, lack of cash for investment, poor transportation links, emigration of skilled Russian workers, and lack of assured markets.

Unit 5

FORM A
Matching
1. c
2. g
3. e
4. f
5. i
6. j
7. k
8. l
9. b
10. d

Fill in the Blank
1. Ivan IV (the Terrible)
2. Turkic
3. Autarky
4. Timur
5. steppe
6. Siberia
7. water
8. Donets Basin
9. Vladimir Lenin
10. Islam

Understanding Ideas
1. a
2. b
3. c
4. b
5. a

Practicing Skills
1. Tajikistan
2. It is geographically closer to Russia than are the other countries.

Composing an Essay
1. Answers will vary but should mention political problems and ethnic conflict. Conflicts over water resources must also be resolved. The countries must continue to clean up the results of Soviet agricultural, industrial, and military practices that damaged the land and water.
2. Answers will vary but should mention that autarky is a system in which a country tries to produce all of the goods it needs. A communist country might want to follow such a policy in order to limit its association with capitalist countries.

FORM B
Short Answer
1. Altay Shan, Pamirs, Tian Shan, Kopet-Dag
2. Much of Russia's rich forest, energy, and mineral resources were wasted because the

government pushed production over conservation. Some of the resources that remain are in remote areas or are of low quality.

3. Answers will vary but should mention that rich oil and gas fields lie between the Volga River and the Ural Mountains. These fields have been crucial to the region's development.
4. Central Asia has relatively few big cities because, throughout most of its history, most of the region's people have been nomads or farmers.
5. Russian traditions that have survived changes in government include religious preferences, an emphasis on education, food and drink choices, and the popularity of dachas.

Practicing Skills

1. Uzbekistan
2. Tajikistan

Composing an Essay

1. Answers will vary but should mention that the Soviets replaced Islamic schools with state-run schools and made public services widely available. Today, literacy rates in the region are well above world averages and life expectancy in most of the region's countries is longer than the world average, particularly for women.
2. Answers will vary but should mention that changes in government have been relatively peaceful after free elections. However, crime and unemployment have increased, making people fearful. The health-care system has disintegrated leaving many elderly with no safety net. Many of those who are getting rich are trying to avoid paying taxes. The middle-classes do not feel secure, fearing that the government may take over business and industry again.
3. Answers will vary but should mention the following factors: outdated equipment, lack of cash for investment, poor transportation links, emigration of skilled Russian workers, and lack of assured markets.
4. Students may say that it was unsuccessful. Without competition from capitalist countries, industrial efficiency and product quality fell. Production of consumer goods and services lagged behind those of capitalist countries. In addition, agricultural production suffered.

Chapter 19

FORM A

Matching

1. d
2. g
3. k
4. h
5. j
6. c
7. e
8. a
9. f
10. b

Fill in the Blank

1. Kabul
2. three
3. Farsi
4. Mecca
5. An Nafūd
6. theocracy
7. Sunni
8. Zagros Mountains
9. Safavids
10. orographic effect

Understanding Ideas

1. a
2. c
3. b
4. b
5. d

Practicing Skills

1. Mecca
2. Persian Gulf

Composing an Essay

1. Answers will vary but should explain that religion is extremely important in the area. Religion is deeply influential in how the country is run and how the economy develops. For example, Orthodox Muslim groups in Afghanistan are limiting women's roles in society. Other countries have theocracies, which means religious law is the law of the government. The agendas of the governments are often based on religious values rather than secular.
2. Answers will vary but should mention that the region is where the world's first major cities were built in the Sumerian civilization. Since that time the area has been conquered many times. The Sumerians were defeated by the Akkadians. The Persians were the next

major civilization to rule the region, followed by Greek and Roman Empires. The Ottoman Empire also ruled for a time. Since the establishment of Islam, the area has become largely Muslim. After the Safavid Persian dynasty, both the Russians and British controlled parts of the region. Today the region is composed of independent countries.

FORM B

Short Answer

1. Two tectonic plates, the African and Arabian, are moving apart. The Red Sea is located between them, so as they move apart it is becoming wider.
2. Arabic culture became prevalent through migration, trade, and the spread of Islam.
3. Saudi Arabia, Bahrain, Kuwait, Oman, Qatar, the United Arab Emirates, Yemen, Iraq, Iran, and Afghanistan
4. The Kurds are an ethnic group found in Iran, Iraq, Syria, and Turkey. They are Muslim but they are not Arabs. Although a distinct ethnic group, they have never had their own country. There is a strong faction that is pushing for self-rule, a situation that is causing political unrest.
5. The most important economic activity in the region is oil and gas production.

Practicing Skills

1. Elburz Mountains
2. Rubʻ Al-Khali

Composing an Essay

1. Answers will vary but should mention that this is the region is where the world's first major cities were built in the Sumerian civilization. Since that time the area has been conquered many times. The Sumerians were defeated by the Akkadians. The Persians were the next major civilization to rule the region, followed by Greek and Roman Empires. The Ottoman Empire also ruled for a time. Since the establishment of Islam, the area has become largely Muslim. After the Safavid Persian dynasty both the Russians and British controlled parts of the region. Today the region is composed of independent countries.
2. Answers will vary but should mention that farming can only take place where water is accessible. Therefore, lands near water or supplied by irrigation are generally used for productive farming.
3. Answers will vary but should mention the large numbers of ethnic and religious factions vying for a limited amount of land and influence. This competition affects the politics and economies of the region. The regions' dependence on one major source of revenue, oil, makes it susceptible to changes in world markets. The dry region and limited water sources make careful management essential to ensure the preservation of the region.
4. Answers will vary but should mention that most developed countries depend on oil to fuel their economies. If the flow of oil from the region were to stop, those developed economies would be affected. Therefore, these countries take great interest in the political situation of the oil-producing countries of Southwest Asia.

Chapter 20

FORM A

Matching

1. l
2. f
3. b
4. c
5. k
6. e
7. h
8. a
9. d
10. g

Fill in the Blank

1. Mediterranean Sea
2. Greek; Turkish
3. Dead Sea
4. Turkey
5. Arabic, Hebrew, and Turkish
6. Ankara
7. Israel
8. Palestine Liberation Organization (PLO)
9. Turkic Muslims
10. Souks

Understanding Ideas

1. c
2. c
3. b
4. b
5. a

Practicing Skills

1. The greatest number of people immigrated to Israel during the 1989–98 period.
2. The large spike in immigration was due to the Holocaust and its aftermath.

Composing an Essay

1. Answers may vary but should mention the 1947 division of Palestine by the United Nations and the establishment of a state for Jews fleeing from the Holocaust. When the British withdrew from the region the next year, the Jewish leadership declared Israel an independent country. Conflicts between Israel and its neighbors in 1956, 1967, 1973, and 1982 resolved little but established the West Bank and Gaza Strips as lands of contention. The establishment of the PLO in 1964 created two rivals for the same land. Jewish settlement in the Golan Heights has also been a source of conflict. Israel does not want to give up these lands as they provide a buffer zone and protection from its neighbors. These lands also allow the Israelis to guard water sources.
2. Answers will vary but should mention that lack of water makes agriculture difficult. Conflicts over water and drought can make life difficult, however. Valuable mineral deposits throughout the region are a source of income for some countries, though others suffer from a lack of resources.

FORM B

Short Answer

1. The major landforms of Turkey are the Pontic and Taurus Mountains and the Anatolian Plain.
2. The PLO stands for the Palestine Liberation Organization. The goal of the PLO is to establish an independent Palestinian state for Arabs.
3. Three important sources of irrigation water in the region are the Jordan, Tigris, and Euphrates Rivers.
4. Water is scarce in the region and therefore a precious commodity. Control of existing water sources such as rivers can cause political problems when one country diverts water sources that are also needed by other countries.
5. Conflicts have arisen over the treatment of Armenian and Kurdish minorities living in Turkey. These groups have complained of unfair treatment. Some want independence. Islamic fundamentalists in Turkey feel that the secular government is too lax while the government has also been criticized for limiting the religious freedoms of Muslims.

Practicing Skills

1. The largest jump in immigration occurred between the 1979–88 and the 1989–98 periods.
2. The number of conflicts that occurred in the region during this period probably discouraged immigration.

Composing an Essay

1. Answers will vary but should mention that Israel wants to keep the area a Jewish homeland while many Arabs want to form an independent Palestinian state. This affects their political viewpoints.
2. Answers will vary but should mention that many cultural practices in the region are based on religion. For example, Muslims of different ethnic groups share a number of customs. Similarly, many legal or business practices in Israel are influenced by the teachings of Judaism.
3. Answers will vary but should mention that Israel has a strong diverse economy while Jordan has a large emigrant Palestinian population, which worsens Jordan's already high unemployment. Jordan is underdeveloped and suffers from a lack of resources, outdated technology, and a weak educational system. Hostile relations have affected both economies.
4. Answers will vary but should mention that struggles between religious factions at odds with each other make it difficult to maintain peace and good relations between countries.

Unit 6

FORM A

Matching

1. c
2. e
3. b
4. g
5. j
6. l
7. d
8. k
9. h
10. i

Fill in the Blank

1. Israel
2. Zagros Mountains
3. Kabul
4. Arabic, Hebrew, and Turkish
5. Ankara
6. Farsi
7. Greek, Turkish
8. orographic effect
9. Mediterranean Sea
10. Safavids

Understanding Ideas

1. c
2. b
3. d
4. b
5. a

Practicing Skills

1. The greatest number of people immigrated to Israel during the 1989–98 period.
2. The large spike in immigration was due to the Holocaust and its aftermath.

Composing an Essay

1. Answers will vary but should mention that the region is where the world's first major cities were built by the Sumerian civilization. Since that time the area has been conquered many times. The Sumerians were defeated by the Akkadians. The Persians were the next major civilization to rule the region, followed by Greek and Roman Empires. The Ottoman Empire also ruled for a time. Since the establishment of Islam, the area has become largely Muslim. After the Safavid Persian dynasty both the Russians and British controlled parts of the region. Today the region is composed of independent countries.
2. Answers will vary but should mention that lack of water makes agriculture difficult. Conflicts over water and drought can make life difficult, however. Valuable mineral deposits throughout the region are a source of income for some countries, though others suffer from a lack of resources.

FORM B

Short Answer

1. The PLO stands for the Palestine Liberation Organization. The goal of the PLO is to establish an independent Palestinian state for Arabs.
2. Three important sources of irrigation water in the region are the Jordan, Tigris, and Euphrates Rivers.
3. Saudi Arabia, Bahrain, Kuwait, Oman, Qatar, the United Arab Emirates, Yemen, Iraq, Iran, and Afghanistan
4. The major landforms of Turkey are the Pontic and Taurus Mountains and the Anatolian Plain.
5. Arabic culture became prevalent through migration, trade, and the spread of Islam.

Practicing Skills

1. The largest jump in immigration occurred between the 1979–88 and the 1989–98 periods.
2. The number of conflicts that occurred in the region during this period probably discouraged immigration.

Composing an Essay

1. Answers will vary but should mention that most developed countries depend on oil to fuel their economies. If the flow of oil from the region were to stop, those developed economies would be affected. Therefore, these countries take great interest in the political situation of the oil-producing countries of Southwest Asia.
2. Answers will vary but should mention that Israel has a strong diverse economy while Jordan has a large emigrant Palestinian population, which worsens Jordan's already high unemployment. Jordan is underdeveloped and suffers from a lack of resources, outdated technology, and a weak educational system. Hostile relations have affected both economies.
3. Answers will vary but should mention that many cultural practices in the region are based on religion. For example, Muslims of different ethnic groups share a

number of customs. Similarly, many legal or business practices in Israel are influenced by the teachings of Judaism.

4. Answers will vary but should mention that farming can only take place where water is accessible. Therefore those lands near water or supplied by irrigation are generally used for productive farming.

Chapter 21

FORM A

Matching

1. i
2. g
3. h
4. b
5. j
6. f
7. k
8. a
9. c
10. l

Fill in the Blank

1. sardines
2. Cairo
3. Red
4. silt
5. Great Britain
6. overcrowding
7. harmattan
8. Tangier
9. Morocco
10. Libya

Understanding Ideas

1. d
2. a
3. a
4. b
5. c

Practicing Skills

1. Tunisia
2. Libya

Composing an Essay

1. Answers may vary but should point out that Islamic fundamentalists think that government should be based strictly on the laws of Islam, whereas the region's governments want to limit the influence of Islam on government.
2. Answers may vary but should mention that the dam is a source of hydroelectric power as well as ample water for irrigation projects. Problems might include a lack of fertile silt in the Nile Delta, because the Nile no longer floods its banks every year. This lack of silt is forcing the use of fertilizers that in turn are polluting the Nile and affecting the fishing industry. The delta is gradually shrinking.

FORM B

Short Answer

1. Egyptians, Phoenicians, Greeks, Romans, Vandals, Byzantines, Ottomans, European countries including Spain, Italy, France, and Great Britain
2. The capitals are Cairo, Egypt; Tripoli, Libya; Tunis, Tunisia; Algiers, Algeria, and Rabat, Morocco.
3. Great Britain wanted the canal because it controlled a major trade route between Europe and Britain's colony in India.
4. Morocco is separated from Europe by only eight miles, across the Strait of Gibraltar. As a result, both European as well as African species of plants and animals are found in Morocco.
5. The major landforms of North Africa include the coastal plains, the Atlas Mountains, the Sahara Desert, and the Nile River Valley.

Practicing Skills

1. Morocco and Tunisia
2. They are small coastal countries that are not dominated by deserts like the other countries are.

Composing an Essay

1. Answers will vary but should mention that many skilled and educated North Africans leave the region to find better jobs in Europe or in oil-rich Arab countries. This could result in a shortage of skilled labor.
2. Answers will vary but should mention that the buildings are tall and close together, and the streets are narrow and twist at odd angles. They were built within the walls of a Casbah and developed this way to deal with the needs of a growing population.
3. Answers will vary but should mention that the natural resources of North Africa include oil, natural gas, iron ore, lead, phosphates, and zinc. There are rich fishing grounds off of Morocco's Atlantic coast. Egypt is an important cotton producer and irrigation has allowed many other crops to be grown in the region.
4. Answers will vary but should mention that

rapid population growth has led to widespread unemployment and overcrowding.

Chapter 22

FORM A

Matching

1. l
2. g
3. b
4. k
5. i
6. a
7. f
8. j
9. e
10. d

Fill in the Blank

1. Rubber; cocoa
2. Poverty
3. primary
4. zonal
5. oil reserves
6. Ghana
7. Mali
8. El Djouf
9. South America
10. market

Understanding Ideas

1. d
2. a
3. d
4. c
5. c

Practicing Skills

1. Korhogo
2. Abidjan

Composing an Essay

1. Answers will vary but should mention planting drought-resistant crops, using irrigation to water crops, planting new trees to replace those cut down, or finding ways to slow the population growth.
2. Answers will vary but each student should choose one of the climate zones of the region and report on weather conditions in that zone.

FORM B

Short Answer

1. West and Central Africa is a region of plains, low hills, and large rivers. There are also a few highland areas and depressions.
2. The slave trade disrupted societies and families. In addition, the guns that Europeans traded for slaves gave coastal forest states an advantage in their wars against interior savanna states.
3. The major resources include tropical timber, good soils, minerals, oil, diamonds, copper, cobalt, cacao, coffee, coconuts, and peanuts.
4. The region's triple heritage is a blend of Islam, European culture, and African culture.
5. Most cities in non-landlocked countries are located on coasts. Most cities were set up as ports and government centers during the colonial era.

Practicing Skills

1. There is only one road to the town and it does not lead to any other cities in the country.
2. Man

Composing an Essay

1. Answers will vary but should mention that the main reason is poverty—parents need their children to work or to care for younger children. Therefore only a small percentage of people in most countries finish high school.
2. Answers will vary but should mention that geographers believe that the upper courses of the rivers date back to when Africa was part of a larger supercontinent, Gondwana, in which the rivers flowed into a large inland lake. They had to find new outlets once the continent broke up.
3. Answers will vary but should mention that societies are based on extended families; families consist of several households; members work together to support the household and care for the old and the young.
4. Answers will vary but should mention that many of the economies of the region were changed from subsistence to market economies. This shift led to changes in the cultures and economic activities of many of the nomadic and herding populations.

Chapter 23

FORM A

Matching

1. d
2. a
3. i
4. l
5. f
6. b
7. k
8. e
9. h
10. j

Fill in the Blank

1. Kush
2. Kenya
3. Djibouti
4. population
5. Sudan
6. Sorghum
7. Blue
8. Addis Ababa
9. Nilotic
10. elevation

Understanding Ideas

1. a
2. a
3. a
4. b
5. b

Practicing Skills

1. Tanzania
2. Uganda and Kenya

Composing an Essay

1. Answers will vary but should mention that while many animals native to the Serengeti Plain are immune, the tsetse fly can spread a deadly disease to livestock. As a result, the area attracts few farmers or herders, which leaves large wild animal populations undisturbed.
2. Answers will vary but should mention that the Europeans helped East Africa by building cities, hospitals, ports, roads, and schools. They harmed East Africa by drawing boundaries that sometimes divided ethnic groups, grouped traditional enemies, or limited groups' access to water.

FORM B

Short Answer

1. the Nilotic peoples, the Cushitic speaking peoples, and the Bantu speaking peoples
2. Tectonic processes beneath Earth's surface lifted the land, causing it to crack and form two rift valleys.
3. Tensions between ethnic groups over land and fair distribution of government aid have led to fighting.
4. Possible answers include gum arabic, beans, corn, rice, sorghum, wheat, coffee, cotton, sugarcane, tea, cloves, and coconut.
5. East African women are often the primary farmers; East African men take care of the livestock.

Practicing Skills

1. Rwanda
2. Somalia, with a 100 percent Muslim population

Composing an Essay

1. Answers will vary but should mention that the central government took over all of the country's agricultural land and urban rental property. It also took over banks, insurance companies, and many other businesses.
2. Answers will vary but should mention that they discovered that the Nile's headwaters are in two different areas. The Blue Nile begins in the highlands of northern Ethiopia. The waters that form the White Nile drain from lake Victoria and through Lake Albert. The Blue Nile and the White Nile join in northern Sudan.
3. Answers will vary but should mention that Swahili's grammar is derived from original languages of the African coast. Over time, many Arabic words were added to the language.
4. Answers will vary but should mention that many people are moving from their farms to the cities in order to find work. Slums have grown up around the cities, and unemployment is common. Political unrest and high crime rates are major issues.

Chapter 24

FORM A

Matching

1. b
2. l
3. e
4. i
5. k
6. g
7. h
8. a
9. d
10. c

Fill in the Blank

1. Drakensberg Range
2. Botswana
3. South Africa
4. Portuguese
5. Nelson Mandela
6. Farming
7. wildlife parks; tropical islands
8. Dutch Reformed
9. copper
10. Orange

Fill in the Blank

1. d
2. c
3. d
4. d
5. b

Practicing Skills

1. Harare
2. Cabinda

Composing an Essay

1. Answers will vary but should include that the country is still struggling because there must be more economic opportunities open to black citizens as well as better education and health care. The increased cost of this equality has strained the economy. Many blacks have become frustrated at the slow pace of change. Difficulties between black ethnic groups have also slowed long-term stability.
2. Answers will vary but should mention that the Dutch came to farm and settle permanently. When the British arrived, the Dutch, or Boers, moved inland and set up two independent republics. They fought Bantu peoples for the possession of this land. When diamonds and gold were discovered, conflict arose between the Boers and the British. The British defeated the Dutch in the Boer war.

FORM B

Short Answer

1. narrow coastal plain, inland plateau, escarpment
2. The high interior plain drops rapidly, creating waterfalls and rapids. This makes boat travel on the river impossible. In addition, sandbars partially block the mouths of the Limpopo and Zambezi rivers, preventing large ships from sailing into the interior.
3. Cities were established by Europeans as seaports and administrative centers. The Dutch built Cape Town and the British built Durban as seaports. Inland cities were developed as mining towns and administrative centers. Johannesburg was originally a mining town. Pretoria and Harare were both administrative cities.
4. Zambia has large copper resources and is one of the world's largest exporters of copper. However, it is very dependent on this one resource and when the price of copper goes down, the economy suffers.
5. poverty, droughts and floods, and disease

Practicing Skills

1. Namib Desert
2. Johannesburg

Composing an Essay

1. Answers will vary but should mention market-oriented farms are larger, often use modern machinery, and produce goods mainly for exports or for cities. Subsistence farmers grow only enough food to feed themselves.
2. Answers will vary but should mention that a cold current flows off the Atlantic coast of southern Africa, creating a low evaporation rate and therefore little precipitation. The only animals that do well in this desert are those that need little water or who can survive on water from fog and dew.
3. Answers will vary but should mention that there is an enormous diversity of languages in each country, which makes communication difficult. European languages are used by governments to allow communication between speakers of different African languages. English is still common in Zimbabwe and Zambia and Portuguese is still the official language of Angola and Mozambique.
4. Answers will vary but should mention that Botswana used to have a small market economy that relied on beef exports. Since the discovery of diamonds in the late 1960s, it has become the world's largest producer of diamonds and now has one of Africa's fastest growing economies.

Unit 7

FORM A

Matching

1. d
2. i
3. f
4. k
5. a
6. l
7. h
8. c
9. e
10. g

Fill in the Blank

1. market
2. Great Britain
3. Kush
4. oil reserves
5. Nilotic
6. Red
7. Morocco
8. zonal
9. Orange
10. Sorghum

Understanding Ideas

1. b
2. b
3. d
4. a
5. b

Practicing Skills

1. Korhogo
2. Abidjan

Composing an Essay

1. Answers will vary but should include the country is still struggling because there must be more economic opportunities open to black citizens as well as better education and health care. The increased cost of this equality has strained the economy. Many blacks have become frustrated at the slow pace of change. Difficulties between black ethnic groups have also slowed long-term stability.
2. Answers may vary but should mention that the dam is a source of hydroelectric power as well as ample water for irrigation projects. Problems might include a lack of fertile soil in the Nile Delta, because the Nile no longer floods its banks every year. This lack of silt is forcing the use of fertilizers that in turn are polluting the Nile and affecting the fishing industry. The delta is gradually shrinking.

FORM B

Short Answer

1. narrow coastal plain, inland plateau, escarpment
2. The major landforms of North Africa are the coastal plains, the Atlas Mountains, the Sahara Desert, and the Nile River Valley.
3. Tensions between ethnic groups over land and fair distribution of government aid have led to fighting.
4. Possible answers include poverty, droughts, floods, and disease.
5. The region's triple heritage is a blend of Islam, European culture, and African culture.

Practicing Skills

1. There is only one road to the town and it does not lead to any other cities in the country.
2. Man

Composing an Essay

1. Answers will vary but should mention market-oriented farms are larger, often use modern machinery and produce goods mainly for exports or for cities. Subsistence farmers grow only enough food to feed themselves.
2. Answers will vary but should mention that rapid population growth has led to widespread unemployment and overcrowding.
3. Answers will vary but should mention that geographers believe that the upper courses of the rivers date back to when Africa was part of a larger supercontinent, Gondwana, in which the rivers flowed into a large inland lake. They had to find a new outlet once the continent broke up.
4. Answers will vary but should mention that Swahili's grammar is derived from original languages of the African coast. Over time, many Arabic words were added to the language.

Chapter 25

FORM A

Matching

1. b
2. e
3. g
4. f
5. j
6. i
7. a
8. h
9. l
10. d

Fill in the Blank

1. Hindi
2. Himalayas
3. Harappan
4. villages
5. Brahmaputra
6. Portuguese
7. infrastructure
8. oil
9. farming
10. New Delhi

Understanding Ideas

1. c
2. a
3. b
4. d
5. a

Practicing Skills

1. 2025
2. Until 2025 they both to appear to be increasing at a similar rate, then China's growth rate is expected to even out while India continues to grow.

Composing an Essay

1. Answers will vary but should note that the Green Revolution was a program designed to allow India to feed itself. It had three main elements: an increase in the amount of cultivated land, two harvests each year, and an increase in production through the use of genetically improved seeds. Irrigation projects were also instituted to overcome farmers' dependence on the monsoon seasons. India can now feed its people, although new programs have proven expensive for small farmers who cannot afford to use pesticides and fertilizers. Irrigation projects have also resulted in environmental disruption and the displacement of peoples so that dams could be built.
2. Answers will vary but should mention that the formal independence movement began with the establishment of the Indian National Congress in 1885. Gandhi led the nonviolent movement for independence through peaceful protest marches and boycotts. Independence was achieved in 1947.

FORM B

Short Answer

1. A low-pressure area over interior Asia causes the summer monsoons, and cold dry winds from Asia's interior cause the winter monsoons.
2. highland, tropical humid, tropical wet and dry, humid subtropical, semiarid, and arid
3. The partition of India was an effort to give both Hindus and Muslims their own states.
4. Cottage industries include silk weaving, figurine carving, silver and gold lacework, and other handicrafts
5. India was a major cotton producer that supplied Britain's new textile industry.

Practicing Skills

1. nearly 1.5 billion
2. India

Composing an Essay

1. Answers will vary but should mention that plant species are tied to rainfall in regions, and so can range from shrubby trees in the Thar Desert to evergreen forests in the Himalayas. Animal life is very varied, with more than 1,200 bird species and nearly 400 snake species found in India. Examples of common animals include crocodiles, deer, elephants, mongooses, monkeys, and tigers.
2. Answers will vary but should mention that more than half of India's land is suitable for farming. Its rivers provide the water necessary to farm the land. The Gangetic Plain stretches across 1,500 miles, and its rich soil is more than 25,000 feet deep in places, which makes farming very productive.
3. Answers should mention the Harappan civilization around the Indus River Valley; the Aryans' movement into the region from central Asia; the Dravidian civilization that was pushed to the Deccan Plateau by the Aryans; the founding of a Muslim kingdom at Delhi around A.D. 1000; the Mughal Empire; Akbar's consolidation of empires; Shāh Jāhan's reign; and Aurangzeb's weak reign and religious intolerance.
4. Answers will vary but should mention that foreign rule angered many Indians who wanted independence. The British had not treated the Indians as equals.

Chapter 26

FORM A

Matching

1. f
2. j
3. a
4. d
5. c
6. b
7. k
8. e
9. l
10. i

Fill in the Blank

1. Bhutan
2. Bhutan
3. Islam
4. East Pakistan
5. Colombo
6. typhoons
7. monsoon
8. low
9. Sri Lanka
10. Nepal

Understanding Ideas

1. a
2. b
3. c
4. a
5. d

Practicing Skills

1. The graph shows two distinct rainy seasons, which would allow for two planting seasons a year.
2. 80°F

Composing an Essay

1. Answers may vary but should mention that across most of the region, a lack of natural resources has hampered industrial development. Rapid population growth and poor living conditions in cities have also hindered growth.
2. Answers may vary but should mention Great Britain divided its colony along religious lines, but this division created a country—Pakistan—composed of two halves separated by hundreds of miles. East Pakistan felt that it had no real power in Pakistan's government, so it broke away and became Bangladesh.

FORM B

Short Answer

1. Bangladesh, Bhutan, Nepal, Sri Lanka, Maldives, Pakistan
2. arid and semiarid in Pakistan; highland in Nepal, Bhutan, and northern Pakistan; tropical climates in Sri Lanka and the Maldives
3. The Hindu Tamil minority is fighting against the Buddhist Sinhalese majority for political independence.
4. Buddhism, Hinduism, and Islam
5. agriculture; some industries; few natural resources; hydroelectric power; tourism

Practicing Skills

1. February
2. There is great variation in the amount of rainfall received over the course of a year but the temperature is fairly constant.

Composing an Essay

1. Answers will vary but should mention that both Bangladesh and Sri Lanka are mostly rural, densely populated, and have closely spaced villages in the fertile farming areas. Bangladesh is much more densely populated than Sri Lanka.
2. Answers will vary but should mention that by partitioning its Indian colony into India and East and West Pakistan, the British divided the colony by religion, but did not consider other factors. East and West Pakistan had nothing else in common, even language. There was constant dissent between them until Bangladesh declared its independence in 1971.
3. Answers will vary but should mention a climate with the unrelenting monsoon season and flooding that severely hampers economic development in Bangladesh; few natural resources, which makes industrialization difficult; and high population growth.
4. Answers will vary but should mention that its lowlands and monsoon flooding have created an ever-changing topography. This instability makes it difficult to attract investors and businesses to the area.

Unit 8

FORM A

Matching

1. a
2. l
3. j
4. d
5. i
6. h
7. g
8. f
9. k
10. c

Fill in the Blank

1. Portuguese
2. low
3. Islam
4. Himalayas
5. New Delhi
6. Nepal
7. farming
8. Sri Lanka
9. Brahmaputra
10. East Pakistan

Understanding Ideas

1. c

2. a
3. d
4. a
5. c

Practicing Skills

1. 2025
2. Until 2025 they both to appear to be increasing at a similar rate, then China's growth rate is expected to even out while India continues to grow.

Composing an Essay

1. Answers may vary but should mention that the diversity of religions has been a source of political conflict in many of the countries, thus slowing down opportunities for economic growth. Some countries have more than one official language, which can prevent a country from being unified in a common goal. Students should mention specific examples from the Indian Perimeter.
2. Answers will vary but should note that the Green Revolution was a program designed to allow India to feed itself. It had three main elements: an increase in the amount of cultivated land, two harvests each year, and an increase in output through the use of genetically improved sees. Irrigation projects were also instituted to overcome farmers' dependence on the monsoon seasons. India can now feed its people, although new programs have proven expensive for small farmers who cannot afford to use pesticides and fertilizers. Irrigation projects have also resulted in environmental disruption and the displacement of peoples so that dams could be built.

FORM B

Short Answer

1. agriculture, some industries; few natural resources; hydroelectric power; tourism.
2. India, as a major cotton producer, could supply Britain's new textile industry.
3. A low-pressure area over interior Asia causes the summer monsoons, and cold dry winds from Asia's interior cause the winter monsoons.
4. Bangladesh, Bhutan, Nepal, Sri Lanka, Maldives, Pakistan
5. highland, tropical humid, tropical wet and dry, humid subtropical, semiarid, and arid

Practicing Skills

1. about 1.5 billion
2. India

Composing an Essay

1. Answers will vary but should mention that more than half of India's land is suitable for farming. Its rivers supply the water needed for agriculture. The Gangetic Plain stretches across 1,500 miles and its rich soil is more than 25,000 feet deep in places, which makes farming very productive.
2. Answers will vary but should mention that its lowlands and monsoon flooding have created an ever-changing topography. This instability makes it difficult to attract investors and businesses to the area.
3. Answers will vary but should mention that by partitioning its Indian colony into India and East and West Pakistan, the British divided the colony by religion, but did not consider other factors. East and West Pakistan had nothing else in common, even language. There was constant dissent between them until Bangladesh declared its independence in 1971.
4. Answers should include the Harappan civilization around the Indus River Valley; the Aryans' movement into the region from central Asia; the Dravidian civilization that was pushed to the Deccan Plateau by the Aryans; the founding of a Muslim kingdom at Delhi about A.D. 1000; the Mughal Empire; Akbar's consolidation of empires; Shāh Jāhan's reign; and Aurangzeb's weak reign and religious intolerance.

Chapter 27

FORM A

Matching

1. i	6. e
2. b	7. k
3. h	8. f
4. g	9. a
5. j	10. c

Fill in the Blank

1. Shang
2. industrialization
3. Tibet
4. Buddhism
5. Mandarin
6. Southern
7. highland
8. Portuguese
9. Mongolia
10. Chiang Kai-shek

Understanding Ideas

1. c
2. a
3. a
4. b
5. b

Practicing Skills

1. 1992–96
2. Answers will vary but students should deduce that the urban sector is growing at a faster rate and so the gap will probably continue to widen.

Composing an Essay

1. Answers will vary but should include the benefits of controlling dangerous flooding along the Chang River, increasing river traffic and trade, and generating hydropower. The drawbacks may include environmental damage such as disruption of ecosystems along the river. Building the dam could also lead to the loss of as much as 240,000 acres of farmland, the loss of more than 1,200 historical sites, and the displacement of 1 to 2 million people.
2. Answers will vary but should mention that Taoism originated in China in the 500s B.C. The followers of Taoism believe there is a natural order to the universe, called Tao. A basic idea of Taoism is to live a simple life in harmony with nature.

FORM B

Short Answer

1. Chinese farmers practice intensive agriculture, which requires a great deal of human labor. Intensive farming methods boost production to high levels.
2. The order of the dynasties is as follows: Shang, Qin, Han, T'ang, Sung. China gets its name from the Qin dynasty.
3. The major rivers are the Huang, Chang or Yangtze, and the Xi.
4. Chinese landscape painting features towering mountains, clouds, and trees. Many Chinese landscapes include descriptive text written in calligraphy.
5. Written Chinese use symbols called characters—some of which are pictograms or simple pictures of the ideas they represent—rather than an alphabet. Chinese writing uses more than 50,000 characters.

Practicing Skills

1. Taiwan's GDP grew from $1,450 per capita in 1978 to $16,500 in 1998, an increase of more than $15,000 over the 20-year period.
2. 1988–93

Composing an Essay

1. Answers will vary but should mention that during the Cultural Revolution, the followers of Mao Zedong tried to rid China of Mao's critics and enemies. Anyone with an education was suspect. Schools and universities were closed. Many old people and scholars were attacked, sent off to labor in the countryside, or killed.
2. Answers will vary but should mention that Deng Xiaoping realized that the Great Leap Forward and the Cultural Revolution had been mistakes. He promoted new policies to modernize China's agriculture, industry, and technology. He also worked to move China toward a market economy.
3. Answers will vary but should mention that market reforms now allow farmers to grow and sell their own crops, build their own private homes, and work in TVEs.
4. Answers will vary. Students should describe Taiwan's high-tech and sports exports, trading partners, and GDP.

Chapter 28

FORM A

Matching

1. d
2. k
3. f
4. c
5. l
6. e
7. j
8. b
9. i
10. h

Fill in the Blank

1. South
2. Chishima Current
3. Sakhalin
4. Japan
5. Kim Jong Il
6. Kyōto
7. subduction
8. hangul
9. surplus
10. Ainu

Understanding Ideas

1. d
2. d
3. b
4. b
5. a

Practicing Skills

1. Hokkaidō
2. P'yŏngyang

Composing an Essay

1. Answers will vary but should mention that Japan lies along a subduction zone and is prey to many earthquakes resulting in tsunamis. Japan is also home to more than 200 volcanoes. In contrast, the Korean peninsula has no active volcanoes and is not prone to earthquakes.
2. Answers will vary but should mention that North Korea has a command economy and is a very poor country. It had been dependent on the Soviet Union for aid, and when that country collapsed it has struggled to stay afloat. South Korea has a export economy and is one of the Asian Tigers. It has a trade surplus with the United States.

FORM B

Short Answer

1. Hokkaidō, Honshū, Shikoku, and Kyūshū
2. The cold Chishima Current from the north cools the summers in the northern part of Japan. The warm Japan Current brings moist marine air to the southern islands, making the summers warm and humid and the winters mild.
3. architecture, ceramics, and painting
4. A strong work ethic, company loyalty, and the keiretsu system allowed Japan's economy to grow rapidly.
5. U.S. forces pushed the Japanese back across the Pacific to Japan, then dropped atomic bombs on Hiroshima and Nagasaki.

Practicing Skills

1. Sea of Japan
2. Kyūshū

Composing an Essay

1. Answers will vary but should mention that after Japan lost control of Korea in 1945, it was occupied by the Soviet Union and the United States. It was divided into North and South Korea at the 38th parallel. Two separate governments were established, a Communist one in North Korea and a democratic one in South Korea, and the occupying forces withdrew. North Korea then invaded South Korea and UN peacekeeping forces were sent in to push North Korea out. An armistice was finally signed and the DMZ now separates the two countries. Relations have been slightly friendlier in recent years.
2. Answers will vary but should mention that both countries have very homogeneous populations. Both countries have largely urban populations. The countries' languages are related, and Buddhism is practiced in both countries, although in Japan it is mixed with Shintoism while in Korea it is blended with Confucianism. There is now a large Christian population in South Korea. Both populations eat rice as their main food. The countries have similar education plans, although fewer Koreans go to college and North Korea places more emphasis on communist ideology in the educational system. Both have a rich heritage in the arts.
3. Answers will vary but should mention that the system was similar to that of medieval Europe, with a shogun as a powerful warlord who ruled over wealthy landlords called *daimyo*. The *daimyo* would then hire samurai to protect their own lands.
4. Answers will vary but should mention that it has helped industry grow by making unified decisions, but now it has gotten too large and has been unable to adapt quickly to growing Asian competition.

Chapter 29

FORM A

Matching

1. c
2. l
3. i
4. f
5. d
6. g
7. h
8. b
9. e
10. j

Fill in the Blank

1. Thailand
2. Khorat Plateau
3. Siam
4. Cambodia
5. shifting
6. Hinduism
7. Hanoi
8. Theravada
9. Laos
10. Laterite

Understanding Ideas

1. a
2. b
3. d
4. a
5. c

Practicing Skills

1. Myanmar
2. Thailand and Vietnam

Composing an Essay

1. Answers will vary but should mention that France established a large colony in the region, called French Indochina. France set up French-language schools and introduced Christianity. Later, after a long bitter war, France let go of the region that would become Vietnam. Cambodia and Laos were also established from what was once French Indochina.
2. Answers will vary but should mention that continued clearing has damaged the habitats of endangered animals such as tigers and elephants and has led to soil erosion and flooding. In addition, a soil known as laterite hardens and becomes useless for agriculture when the forest cover is removed.

FORM B

Short Answer

1. The British and the French set up plantations for growing export crops, built roads and railroads, set up English- and French-language schools, and introduced Christianity.
2. Mainland Southeast Asia's four major rivers are the Irrawaddy, the Chao Phraya, the Mekong, and the Hong (Red) River. Fertile alluvial soils support intensive farming.
3. It is the region's largest freshwater lake. During the dry season, a river flows south from the lake into the Mekong River. However, during the wet season, the Mekong's water level is higher than the lake's and the flow is reversed. Water from the Mekong is pushed upstream into the lake, more than doubling its normal size.
4. The Burmese speak a Sino-Tibetan language related to Chinese. Languages of the Tai family are spoken in Thailand and Laos. The Vietnamese and Khmer peoples speak languages from the Austro-Asiatic family.
5. During the wet monsoon, severe flooding is common. During the dry season, on the other hand, fires can be a major problem.

Practicing Skills

1. Laos
2. Thailand

Composing an Essay

1. Answers will vary but should mention that most of the region's population is Buddhist. Thai men often spend time working and serving in monasteries, and Buddhist men in Laos have traditionally been expected to become monks for a while. The countries celebrate Buddhist festivals and holidays.
2. Answers will vary but should mention that the Khmer were a highly developed culture group in the region. They dominated Cambodia from the A.D. 800s to the end of the 1100s. Angkor Wat is a temple complex built by the Khmer and is a testament to their advanced civilization.
3. Answers will vary but should mention that the United States wanted to stop the spread of communism in Asia. American policy was based on the domino theory, which stated that if one of Southeast Asia's countries became Communist, then the others would follow.

4. Answers will vary but should mention that many rural people hold to traditional cultural practices, including religious festivals and local animist beliefs. Rural peoples are more likely to wear traditional garments of the region, while Western-style clothing is common in urban areas, as is American fast food.

Chapter 30

FORM A

Matching

1. f
2. j
3. l
4. a
5. c
6. g
7. d
8. e
9. h
10. b

Fill in the Blank

1. tourism
2. the Philippines
3. Batik
4. Singapore
5. cities
6. wet-rice or paddy
7. Jakarta
8. Borneo
9. Spain
10. Brunei

Understanding Ideas

1. a
2. a
3. c
4. b
5. a

Practicing Skills

1. approximately 100 inches
2. Temperature stays the same virtually year round. Rainfall is fairly steady for nine months followed by a three-month rainy season.

Composing an Essay

1. Answers may vary but should mention that a constant warm climate allows for tourism year round; its location between Asia and the United States provides access to two large markets; valuable oil and gas reserves create cash flow; and the climate is conducive to a strong agricultural economy.
2. Answers may vary but should mention that ethnic and religious differences have led to political strife and demands for independence, which adversely affects the economy.

FORM B

Short Answer

1. The climates and biomes of the region are influenced by the region's location in the tropics, by the effects of the monsoon flow, and by the higher elevations of the region's mountains.
2. Singapore is located on a major shipping route. In addition, many financial and high-technology companies have opened offices there.
3. The Portuguese came in search of spices, the Spaniards wanted to Christianize and colonize the islands, the Dutch wanted to control the spice and tea trade, and the British wanted to colonize and trade.
4. Tectonic processes, particularly volcanic activity, have shaped the region's landforms.
5. In wet-rice cultivation, farmers construct dikes to create rice paddies in lowland areas or mud terraces in hilly areas; in dry-rice cultivation, they plow fields and plant rice seeds; and in slash-and-burn agriculture, they cut small areas of forest and burn fallen trees to create clearings in which they can farm.

Practicing Skills

1. July
2. Answers should mention constant warm temperatures, high rainfall, and a rainy season, all of which are characteristic of a tropical humid climate.

Composing an Essay

1. Answers will vary but should mention that these Pacific Rim countries have experienced rapid industrialization and great economic growth in recent decades.
2. Answers will vary but should mention that Singapore and Brunei are similar in that they are the two wealthiest, most prosperous, and most urbanized countries in the region. How that wealth is attained is very different. Brunei is almost completely dependent on the export of a single commodity, oil, while Singapore has an extremely diverse economy in business, finance, and high-technology companies. Both countries were European colonies, but Brunei is now ruled by a sultan and is

predominantly Muslim while Singapore is known for its strict and rigid laws. Its population is much more diverse than is Brunei's.

3. Answers will vary but should mention that the tropical rain forests found in island Southeast Asia house the greatest variety of plant and animal life on Earth. However, much of this forest is being lost to human activities such as logging, farming, and development. The loss of animal habitats to these human activities endangers the animals.
4. Answers will vary but should mention that the Philippines is the most homogeneous country in the region—more than 90 percent of the people are ethnic Malays. In addition, Christians make up more than 90 percent of the population, a much larger percentage than in any other of the region's countries.

Unit 9

FORM A

Matching

1. h
2. g
3. a
4. b
5. f
6. k
7. c
8. i
9. l
10. e

Fill in the Blank

1. Kim Jong Il
2. Cambodia
3. Shang
4. Ainu
5. the Philippines
6. highland
7. Singapore
8. Tibet
9. the Spanish
10. Laos

Understanding Ideas

1. a
2. a
3. d
4. b
5. b

Practicing Skills

1. approximately 100 inches
2. Temperature stays the same virtually year round. Rainfall is fairly steady for nine months followed by a three-month rainy season.

Composing an Essay

1. Answers may vary but should mention that ethnic and religious differences have led to political strife and demands for independence, which adversely affects the economy.
2. Answers will vary but should mention that North Korea has a command economy and is a very poor country. It had been dependent on the Soviet Union for aid, and when that country collapsed it has struggled to stay afloat. South Korea has a export economy and is one of the Asian Tigers. It has a trade surplus with the United States.

FORM B

Short Answer

1. The climates and biomes of the region are influenced by the region's location in the tropics, by the effects of the monsoon flow, and by the higher elevations of the region's mountains.
2. Chinese farmers practice intensive agriculture, which requires a great deal of human labor. Intensive farming methods boost production to high levels.
3. The British and the French set up plantations for growing export crops, built roads and railroads, set up English- and French-language schools, and introduced Christianity.
4. In wet-rice cultivation, farmers construct dikes to create rice paddies in lowland areas or mud terraces in hilly areas; in dry-rice cultivation, they plow fields and plant rice seeds, and in slash-and-burn agriculture, they cut small areas of forest and burn fallen trees to create clearings in which they can farm.
5. The cold Chishima Current from the north cools the summers in the northern part of Japan. The warm Japan Current brings moist marine air to the southern islands, making the summers warm and humid and the winters mild.

Practicing Skills

1. Taiwan's GDP grew from $1,450 per capita in 1978 to $16,500 in 1998, an increase of more than $15,000 over the 20-year period.

2. 1988–93

Composing an Essay

1. Answers will vary but should mention that both countries have very homogeneous populations. Both countries have largely urban populations. The countries' languages are related, and Buddhism is practiced in both countries, although in Japan it is mixed with Shintoism while in Korea it is mixed with Confucianism. There is now a large Christian population in South Korea. Both populations eat rice as their main food. The countries have similar education plans, although fewer Koreans go to college and North Korea places more emphasis on communist ideology in the educational system. Both have a rich heritage in the arts.
2. Answers will vary but should mention that Thai men often spend time working and serving in monasteries, and Buddhist men in Laos have traditionally been expected to become monks for a while. The region's countries celebrate Buddhist festivals and holidays.
3. Answers will vary but should mention that these Pacific Rim countries have experienced rapid industrialization and great economic growth in recent decades.
4. Answers will vary but should mention that during the Cultural Revolution, the followers of Mao Zedong tried to rid China of Mao's critics and enemies. Anyone with an education was suspect. Schools and universities were closed. Many old people and scholars were attacked, sent off to labor in the countryside, or killed.

Chapter 31

FORM A

Matching

1. f
2. e
3. c
4. h
5. l
6. j
7. a
8. k
9. b
10. i

Fill in the Blank

1. sheep; cattle
2. bats
3. Great Dividing Range
4. North
5. Dreamtime
6. Cook Strait
7. rain forests
8. Great Britain
9. raw
10. Pacific Ring of Fire

Understanding Ideas

1. b
2. d
3. d
4. c
5. b

Practicing Skills

1. Australia's population grew along the coasts, with little population in the interior due to the harsh climate and physical conditions of the interior.
2. The names of the territories reveal its roots as colony of Great Britain: New South Wales is perhaps named after the country of Wales in Great Britain, Queensland after Queen Victoria who ruled Great Britain during the later half of the 1800s and so on. Victoria is also named after Queen Victoria.

Composing an Essay

1. Answers may vary but should mention that, unlike New Zealand, Australia was established as a prison colony. Both countries were colonized by Great Britain, and their populations remain predominantly European with English as the official language. Europeans in New Zealand attempted to make treaties with the indigenous Maori, although violence erupted between the two groups. Australian colonists drove Aborigines from their land. Many of the Aborigines died of European diseases.
2. Answers may vary but should mention that the creation of the EU has forced both countries to seek new markets for their products. Both have turned increasingly to Asia and the United States for trade.

FORM B

Short Answer

1. Much of Australia is between about 20°

and 30° south latitude, an area dominated by a warm subtropical high-pressure zone with dry air. Another factor is Australia's generally low elevation. There are few mountains tall enough to create an orographic effect.

2. Most Australians have settled in the southeast, which has pleasant climates and reliable rainfall. Few people live in the dry interior, where a lack of water makes settlement risky.

3. Mining, agriculture, manufacturing, international trade, and tourism are important to Australia's economy.

4. Tectonic processes have created active volcanoes, geysers, and hot springs in New Zealand. They also produce earthquakes.

5. European settlers in New Zealand began taking Maori lands and introduced diseases that killed many Maori.

Practicing Skills

1. Based on the settlement patterns show in the map and the fact that Western Australia has not been divided into multiple states, the terrain and climate of Western Australia are probably severe. The land may not be suitable for agriculture or the climate may not produce a high amount of rainfall.

2. five states, plus the Australian Capital Territory

Composing an Essay

1. Answers will vary but should mention that some of New Zealand's factories make processed foods like butter and cheese. Other industries include wood and paper production and textiles.

2. Answers will vary but should mention Australia's dependence on manufactured imports while it exports only raw materials. The government is trying to encourage the development of industry to combat this weakness.

3. Answers will vary but should mention that Australia was originally settled as a prison colony. Later people came to farm and raise sheep. The discovery of gold in 1851 attracted more settlers. Eventually six large colonies developed. New Zealand's first European colonists came from Australia. They came as missionaries, traders, and whalers rather than farmers. However, New Zealand also developed a strong agricultural and sheep herding economy.

4. Answers will vary but should mention that both islands are mountainous and forested. Both have a marine west coast climate. However, South Island is larger and has higher elevations than North Island. Three fourths of New Zealand's population lives on North Island.

Chapter 32

FORM A

Matching

1. b
2. g
3. h
4. c
5. k
6. d
7. j
8. f
9. a
10. l

Fill in the Blank

1. Low
2. fish
3. Oceanic
4. high
5. Melanesia
6. Whale oil
7. Papua New Guinea
8. Guam
9. France
10. Nauru

Understanding Ideas

1. c
2. b
3. b
4. d
5. a

Practicing Skills

1. C
2. A

Composing an Essay

1. Answers will vary but should mention that the cultures of the Pacific represent a blend of elements from the various peoples who migrated into the region. Students should also note the effect of World War II on the region and the subsequent creation of trust territories. Europeans have had the biggest impact on the area, changing the basic subsistence economy to a more market-oriented economy. Trade has allowed new ideas and products to enter the cultures of the region.

2. Answers will vary but should mention the sheer size of the area this region covers. The large distances between islands make communication and trade more difficult. The fact that each country is generally made up of a number of small islands with little land mass makes development difficult. The huge diversity in populations and cultures also makes it difficult for a unified economy to develop.

FORM B

Short Answer

1. Pidgin languages are simplified languages based on English. They are useful in that people who speak different languages are able to communicate.
2. EEZs allow countries to charge fees for economic activities within their boundaries. These fees from foreign businesses provide much needed income.
3. The region is concerned with global warming as the melting of polar ice caps could catastrophically affect low-lying islands. The long-term effects of nuclear testing in the area are also a concern.
4. Micronesia, Melanesia, and Polynesia
5. Tropical climates with high rainfall are most common in the Pacific Islands. Papua New Guinea does have areas with a highland climate high in its mountainous interior.

Practicing Skills

1. F
2. D

Composing an Essay

1. Answers will vary but should note that high islands are rocky and mountainous. They either sit on continental shelves or represent the peaks of undersea mountains. Low islands form from coral and are usually small and flat. Low islands are less able to sustain human populations because they lack freshwater.
2. Answers will vary but should mention that the high population increase has forced many islanders to migrate to the cities where many cannot find jobs. Many people are also moving out of the region. These are often skilled workers, so many islands suffer from a lack of skilled labor.
3. Answers will vary but should mention that coral reefs contain approximately 25 percent of the plant life in the ocean and are necessary to maintain diversity.
4. Answers will vary but should mention that populations probably first migrated from Southeast Asia. Students should also discuss colonization by European countries and its subsequent impact.

Unit 10

FORM A

Matching

1. j
2. e
3. c
4. k
5. l
6. d
7. f
8. g
9. b
10. a

Fill in the Blank

1. extensive
2. high
3. Great Dividing Range
4. Whale oil
5. Dreamtime
6. France
7. trade winds
8. raw
9. Melanesia
10. Pacific Ring of Fire

Understanding Ideas

1. a
2. c
3. d
4. a
5. b

Practicing Skills

1. Australia's population grew along its coasts, with little population in the interior due to the harsh climate and physical conditions of the interior.
2. The names of the territories reveal its roots as colony of Great Britain: New South Wales is perhaps named after the country of Wales in Great Britain, Queensland after Queen Victoria who ruled Great Britain during the later half of the 1800s, and so on. Victoria is also named after Queen Victoria.

Composing an Essay

1. Answers will vary but should mention that the cultures of the Pacific represent a

blend of elements from the various peoples who migrated into the region. Students should also note the effect of World War II on the region and the subsequent creation of trust territories. Europeans have had the biggest impact on the area, changing the basic subsistence economy to a more market oriented economy. Trade has allowed new ideas and products to enter the cultures of the region.

2. Answers may vary but should mention that the creation of the EU has forced both countries to seek new markets for their products. Both have turned increasingly to Asia and the United States for trade.

FORM B

Short Answer

1. Much of Australia is between about 20° and 30° south latitude, an area dominated by a warm subtropical high-pressure zone with dry air. Another factor is Australia's generally low elevation. There are few mountains tall enough to create an orographic effect.
2. Melanesia, Polynesia, and Micronesia
3. Mining, agriculture, manufacturing, international trade, and tourism are important to Australia's economy.
4. Pidgin languages are simplified languages based on English. They are useful in that people who speak different languages are able to communicate.
5. Tectonic processes have created active volcanoes, geysers, and hot springs in New Zealand. They also produce earthquakes.

Practicing Skills

1. Based on the settlement patterns show in the map and the fact that Western Australia has not been divided into multiple states, the terrain and climate of Western Australia are probably severe. The land may not be suitable for agriculture or the climate may not produce a high amount of rainfall.
2. five states, plus the Australian Capital Territory

Composing an Essay

1. Answers will vary but should mention that both islands are mountainous and forested. Both have a marine west coast climate. However, South Island is larger and has higher elevations than North Island. Three fourths of New Zealand's population lives on North Island.
2. Answers will vary but should mention that coral reefs contain approximately 25 percent of the plant life in the ocean and are necessary to maintain diversity.
3. Answers will vary but should mention that Australia was originally settled as a prison colony. Later people came to farm and raise sheep. The discovery of gold in 1851 attracted more settlers. Eventually six large colonies developed. New Zealand's first European colonists came from Australia. They came as missionaries, traders, and whalers rather than farmers. However, New Zealand also developed a strong agricultural and sheep herding economy. Both were British colonies.
4. Answers will vary but should note that high islands are rocky and mountainous. They either sit on continental shelves or represent the peaks of undersea mountains. Low islands form from coral and are usually small and flat. Low islands are less able to sustain human populations because they lack freshwater.